WORLD MODELLING

ARCHITECTURAL MODELS IN THE 21ST CENTURY

Guest-edited by
MARK MORRIS
AND MIKE ALING

03 | Vol 91 | 2021

WORLDMODELLING 03/2021

About the Guest Editors

Mark Morris and Mike Aling
05

Introduction
Scaling Up
The Many Worlds of the Architectural Model

Mark Morris and Mike Aling
06

More on the Model
Building on the Ruins of Representation

Christian Hubert
14

Miniature Places for Vicarious Visits
Worldbuilding and Architectural Models

Mark JP Wolf
22

Polyphonic Dreams
Storytime in Synthetic Reality

Kate Davies
32

Studio Klarenbeek & Dros, Mycelium Chair, 2013

Worlds Without End

Mark Cousins
40

Remodelling
Home as Cosmos

Chad Randl
56

Handmade Worlds
Constructing an Inhabitable Modelscape

Pascal Bronner and Thomas Hillier
48

Dennis Maher, City Chancel, 2017

Everything You See is Yours
Step Towards the Certainty of Uncertainty

Theodore Spyropoulos
64

ISSN 0003-8504
ISBN 978 1119 747222

Guest-edited by **Mark Morris and Mike Aling**

Model & Fragment
On the Performance of Incomplete Architectures

Thea Brejzek and Lawrence Wallen

74

Models as Objects
The Installation as Architectural Encounter

James A Craig and Matt Ozga-Lawn

82

Sofia Belenky, Null Island, Intermediate 5 unit, Architectural Association, London, 2017

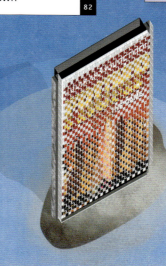

Zero Zero Ze(r)ro(r)
How the Cartographic Thirst to Project the Real Reveals Spaces for the Creation of New Worlds

Ryan Dillon

88

The White Cube in Virtual Reality

Kathy Battista

102

From Mimicry to Coupling
Some Differences, Challenges and Opportunities of Bio-Hybrid Architectures

Phil Ayres

96

Backgarden Worldbuilding
The Architecture of the Model Village

Mike Aling

112

Paracosmic Project
The Architectural Long Game

Mark Morris

120

From Another Perspective
A Surrealist Rococo Master Kris Kuksi

Neil Spiller

128

Sarah Meyohas, Rope Speculation, 2015

Contributors

134

3

Editorial Offices
John Wiley & Sons
9600 Garsington Road
Oxford
OX4 2DQ

T +44 (0)1865 776 868

Editor
Neil Spiller

Managing Editor
Caroline Ellerby
Caroline Ellerby Publishing

Freelance Contributing Editor
Abigail Grater

Publisher
Todd Green

Art Direction + Design
CHK Design:
Christian Küsters
Barbara Nassisi

Production Editor
Elizabeth Gongde

Prepress
Artmedia, London

Printed in the United Kingdom
by Hobbs the Printers Ltd

Denise Bratton
Paul Brislin
Mark Burry
Helen Castle
Nigel Coates
Peter Cook
Kate Goodwin
Edwin Heathcote
Brian McGrath
Jayne Merkel
Peter Murray
Mark Robbins
Deborah Saunt
Patrik Schumacher
Ken Yeang

EDITORIAL BOARD

Journal Customer Services
For ordering information, claims and any enquiry concerning your journal subscription please go to www.wileycustomerhelp.com/ask or contact your nearest office.

Americas
E: cs-journals@wiley.com
T: +1 877 762 2974

Europe, Middle East and Africa
E: cs-journals@wiley.com
T: +44 (0)1865 778315

Asia Pacific
E: cs-journals@wiley.com
T: +65 6511 8000

Japan (for Japanese-speaking support)
E: cs-japan@wiley.com
T: +65 6511 8010

Visit our Online Customer Help available in 7 languages at www.wileycustomerhelp.com/ask

Print ISSN: 0003-8504
Online ISSN: 1554-2769

Prices are for six issues and include postage and handling charges. Individual-rate subscriptions must be paid by personal cheque or credit card. Individual-rate subscriptions may not be resold or used as library copies.

All prices are subject to change without notice.

Identification Statement
Periodicals Postage paid at Rahway, NJ 07065. Air freight and mailing in the USA by Mercury Media Processing, 1850 Elizabeth Avenue, Suite C, Rahway, NJ 07065, USA.

USA Postmaster
Please send address changes to *Architectural Design*, John Wiley & Sons Inc., c/o The Sheridan Press, PO Box 465, Hanover, PA 17331, USA

Rights and Permissions
Requests to the Publisher should be addressed to:
Permissions Department
John Wiley & Sons Ltd
The Atrium
Southern Gate
Chichester
West Sussex PO19 8SQ
UK

F: +44 (0)1243 770 620
E: Permissions@wiley.com

All Rights Reserved. No part of this publication may be reproduced, stored in a retrieval system or transmitted in any form or by any means, electronic, mechanical, photocopying, recording, scanning or otherwise, except under the terms of the Copyright, Designs and Patents Act 1988 or under the terms of a licence issued by the Copyright Licensing Agency Ltd, 5th Floor, Shackleton House, Battle Bridge Lane, London SE1 2HX, without the permission in writing of the Publisher.

Subscribe to ⌀
⌀ is published bimonthly and is available to purchase on both a subscription basis and as individual volumes at the following prices.

Prices
Individual copies:
£29.99 / US$45.00
Individual issues on
⌀ App for iPad:
£9.99 / US$13.99
Mailing fees for print may apply

Annual Subscription Rates
Student: £93 / US$147
print only
Personal: £146 / US$229
print and iPad access
Institutional: £346 / US$646
print or online
Institutional: £433 / US$808
combined print and online
6-issue subscription on
⌀ App for iPad: £44.99 /
US$64.99

Front cover: Minimaforms (Theodore and Stephen Spyropoulos), *Of and In The World*, London, 2017. Photo Theodore Spyropoulos

Inside front cover: Ezgi Terzioglu, Crossing the Line, Intermediate Unit 5, Architectural Association (AA), London, 2018. © Architectural Association, School of Architecture

Page 1: Stasus (James A Craig and Matt Ozga-Lawn), *Everest Death Zone*, 2013. © Stasus 2009–16

03/2021

⌀ ARCHITECTURAL DESIGN
May/June 2021
Profile No. 271

Disclaimer
The Publisher and Editors cannot be held responsible for errors or any consequences arising from the use of information contained in this journal; the views and opinions expressed do not necessarily reflect those of the Publisher and Editors, neither does the publication of advertisements constitute any endorsement by the Publisher and Editors of the products advertised.

ABOUT THE
GUEST-EDITORS

MARK MORRIS AND MIKE ALING

Mark Morris and Mike Aling have been discussing their shared fascination with architectural models, and their continued development in light of new media technologies, for a number of years. Both have looked to how modelling has evolved within architectural education and practice. This issue of △, conceived in a pub opposite a model village, is their first joint project on the subject.

Mark Morris is Head of Teaching at the Architectural Association (AA) in London where he lectures in history and theory. He chairs the AA's Teaching and Learning Committee and is a member of its Academic Board and Senior Management Team. He studied architecture at Ohio State University where he received the AIA Henry Adams medal, and completed his PhD at the University of London's Consortium doctoral programme supported by the RIBA Research Trust Award. He was previously Director of Graduate Studies in the Field of Architecture and, subsequently, Director of Exhibitions at Cornell University's College of Architecture, Art and Planning in Ithaca, New York. He is the author of *Models: Architecture and the Miniature* (Wiley, 2006) and *Automatic Architecture* (University of North Carolina, 2007), as well as chapters in many other books. His essays have also featured in *Domus*, *Log*, *Frieze*, △, *Cabinet*, the *Cornell Journal of Architecture* and *Critical Quarterly*. He is a member of the Victoria and Albert Museum's Architectural Models Network and RIBA Academic Publications Panel. His research focuses on questions of visual representation, architecture in fiction, and the study of paracosms as mental places for detailed creative work of long duration.

Mike Aling is a senior lecturer at the University of Greenwich School of Design in London, where he is the MArch Architecture programme leader. He runs Unit 14 at Greenwich, a postgraduate design group that explores new architectural modelling systems, processes and languages. In recent years Unit 14 has increasingly looked to worldbuilding as a methodology to propose architectural projects through the vehicle of models. His research examines and speculates on the continuing evolution of digital architectural modelling methodologies and procedures, and the consequences of this development on the histories and theories of the architectural model. He is fascinated by model villages and their affordances to architecture, and is currently undertaking a PhD at Newcastle University on this subject. His other ongoing research interests explore the history and potential future of the architectural book, and the role of architecture in wider printed and digital media design publishing frameworks. He has been published in a number of architectural books and journals, including △. His design work has been exhibited in the US, South Korea, Austria and London. △

Scaling Up

The Many Worlds of the Architectural Model

INTRODUCTION
MARK MORRIS AND MIKE ALING

Greg Lynn, *Satellite Worlds*, 'Other Space Odysseys: Greg Lynn, Michael Maltzan, Alessandro Poli', Canadian Centre for Architecture (CCA), Montreal, Quebec, 2010

Partly a recollection of the moon landing, partly a sci-fi design brief, Lynn's elliptical satellite worlds exhibited at the CCA are microcosms freed of earthly constraints, but also in dialogue with Earth.

The meaning of 'model' can range from the loftiest notions of paragons and ideals through to the practical operations of gluing cardboard into a 3D form, through to the most complex of digital constructions. The spectrum of what might constitute an architectural model continues to stretch and to be redefined. It is increasingly difficult to pigeonhole the architectural model as a singular object or method in the 21st century. With the ubiquity of building information modelling (BIM) in the profession, the processes of modelling and the construction of models are now arguably the dominant mode of production. We are reaching a point where every building is born a digital model, concretised in the phenomenal world over time like the slow setting of a cast. The advancement and availability of 3D-modelling software has also allowed designers to be ever more ambitious with their models, to the point where entire imagined and digitally constructed worlds can thrive. This issue of ⌂ does not attempt to clarify what an architectural model might be today; it aims to discuss a new shift that involves how the model sits in a world of its own making – as a 'worldmodel'.

Recent History

Writing on the architectural model had its 'boom' moment in the mid-2000s. These works often opened by clarifying their engagement with this most slippery of terms. In both Albert C Smith's *Architectural Model as Machine: A New View of Models from Antiquity to the Present Day* (2004)[1] and Karen Moon's *Modeling Messages: The Architect and the Model* (2005),[2] for example, the discussion is pinned firmly to the architect's physical model. Many of the writings were directly inspired by the preceding 'Idea as Model' exhibition of 1976; however, this exhibition took a decidedly different tack. The exhibition was curated by Peter Eisenman and presented physical model works from a large number of the US architectural vanguard of the time at the Institute for Architecture and Urban Studies (IAUS) in New York, with the aim of promoting the architectural model as capable of more than communicating proposals to clients in miniaturised (and often simplified) form. These artefacts were intended to become works in their own right, beyond the representational, as 'conceptual models'. However, in the delayed exhibition publication of 1981, Christian Hubert declared that this intention was always flawed, and that the autonomy of the architectural model was an impossibility at the time due to its intractable relationship with the subject/proposal that it represents.[3] The 'Idea as Model' project was a ready response to Arthur Drexler's 1975–6 'The Architecture of the École des Beaux-Arts' exhibition at the Museum of Modern Art (MoMA) in New York, which, naturally, glorified the drawing. Both shows contributed to postmodern architecture, the scrappier one punching well above its weight.

In the decades that followed, model interest swung towards establishing their provenance and preserving them as objects of study. Dean of the Center for Advanced Study in the Visual Arts (CASVA) of the National Gallery in Washington DC, Henry A Millon's exhibitions and catalogues on Renaissance and Baroque architecture from the mid-1990s included extensive scholarship around the few extant models from those eras. Tellingly, these works were published before the near-ubiquity of the digital model, perhaps to reassert the importance of the physical model in a time when it was evaporating into pixels. There were of course exceptions to the model-as-physical rule when writing on the model in the 2000s. In *The Model and its Architecture* (2008),[4] for example, Patrick Healy opened up the term into its broadest sense, moving from Plato to Deleuze and back again. Today architects have come to accept 3D-printed and other digitally fabricated models as standard outputs, often forgetting their one-time novelty.

Greg Lynn,
New City,
'Other Space Odysseys:
Greg Lynn, Michael Maltzan,
Alessandro Poli',
Canadian Centre for
Architecture (CCA),
Montreal, Quebec,
2010

Model of one of a series of postulate megacities or small worlds that loop and coil around themselves.
This particular world, a single continuous city, is populated through and mediated by social media. The model hovers above a mirror installed in the plinth below.

Worldbuilding and Worldmaking

Worldbuilding is the practice of constructing imaginary worlds. Long associated with fantasy epics and para-literatures, the concept has burgeoned into a field that has also encompassed media studies, film and cinema, video games studies, urbanism, landscape and, of course, architecture. Mark JP Wolf has written extensively on the subject and is joining the debate in this issue. Otherwise known as 'subcreation' or 'conworlding' (constructed worlds), the success of these fabrications relies on their consistent upholding of self-instigated internal rules and logic systems. Architecture has a complex relationship with worldbuilding: architects often imagine a slightly newer version of our current world, while being intrinsically tied to its realities. It is this propensity to imagine new world spaces not necessarily tied to the actual existing world that we increasingly see in the making of architectural models.

There is the danger that worldbuilding becomes a catch-all term for any imagined worldspace. And while there is no discernible difference between 'worldbuilding' and 'world-building', the term 'worldmaking', however, is distinct. Worldmaking is discussed in philosopher Nelson Goodman's *Ways of Worldmaking* (1978)[5] as a form of treatise on how ideas affect the production of the world (in all of its guises, not solely the haptic and tactile). Architects arguably have a close association to Goodman's notion: they are in the business of producing design imaginaries that act as catalysts for change in our actual (but not necessarily phenomenal) world(s).

Denis Maher,
City Wall-scape,
Fargo House,
Buffalo, New York,
2014

An interior of Dennis Maher's magnum opus, Fargo House. Maher purchased the abandoned house for $10,000 in 2009. Rather than renovate, he interrogated what was left, reconfigured and exposed the structure, and in-filled certain gaps with his packrat collection of architectural fragments and architectural toys. This room includes *City Wall-scape*, a melange of donated dolls' houses and scale models, something reminiscent of certain moments in Sir John Soane's house in London.

James Lawton,
The Gamification of Alt-Erlaa,
Vienna,
MArch Architecture,
Unit 14,
University of Greenwich
School of Design,
London,
2018

Student projects may signpost the way for the architectural model as a methodology for imagined new versions of the world, with physical models often increasing in size and scope, perhaps partly due to streamlined digital fabrication workflow. In this project, Vienna's social housing megalith Alt-Erlaa is reimagined in a world of ubiquitous social credit, becoming a playground where all mundane activities are gamified in order to increase the wellbeing of the inhabitants.

Isobel Eaton,
Hotel Hypnagogia,
MArch Architecture,
Unit 14,
University of Greenwich
School of Design,
London,
2019

Through the simple act of inviting us into the large immersive interior of the model (we enter by lying on our back and sliding in on a wheeled bed-board), the viewer inhabits an alternative world where we live in a state of hypnagogia, the space between wake and sleep. The hotel encourages occupants to delight in the experience of drifting off. With no assigned rooms, beds or set accommodation, guests are induced to sleep where they settle in a series of controlled communal environments – as is the viewer, who is invited to sleep in the comfort of the model.

Scales of Inquiry

This issue of ⌂ was largely written during the pandemic lockdown. Rather than that circumstance being a hindrance, it seemed to focus the minds of contributors and influence some thinking on worldmodelling. Christian Hubert (pp 14–21) explicitly makes this case, finding the timing of the development and thinking around worldmodels aligned to growing awareness around the fragile ecological state of the planet. Climate change, political turmoil and social upheaval prompt a turning towards worldmodelling as a coping mechanism, both as escapism and a speculative space for testing ways to heal the planet. Hubert puts his case in the context of his earlier observations of models asserted in his influential 'Ruins of Representation' essay featured in the 1981 *Idea as Model* catalogue.[6] In that 40-year gap, Hubert finds a proliferation of worldmodel thinking.

Mark JP Wolf is arguably the world's leading scholar on the subject of worldbuilding. In his article (pp 23–31) he returns to his earlier passion for architecture, and discusses the many ways in which the architectural model has, and continues to, operate as a worldbuilding device. The appeal of miniature architectures, both actual and virtual, is discussed in relation to his theories on worldbuilding.

Co-founder of Unknown Fields, Kate Davies looks to how worldmodels or slices thereof can be expressed in different media (pp 32–9). Film and video, she claims, are especially adept at communicating not just the look of alternative worlds, but their reason for being. Atmosphere, datascapes, duration and narrative make such models seem alive.

In his last penned essay (pp 40–47), the late theorist Mark Cousins points out the telling differences between *a* world and *the* world, and the useful creative friction between the two. His unpacking of the terminology of worldmodelling brings out salient arguments and conceptual minefields. The endeavour reveals the tenebrous hold we have of any world, and how a world-as-construct is so easily deconstructed.

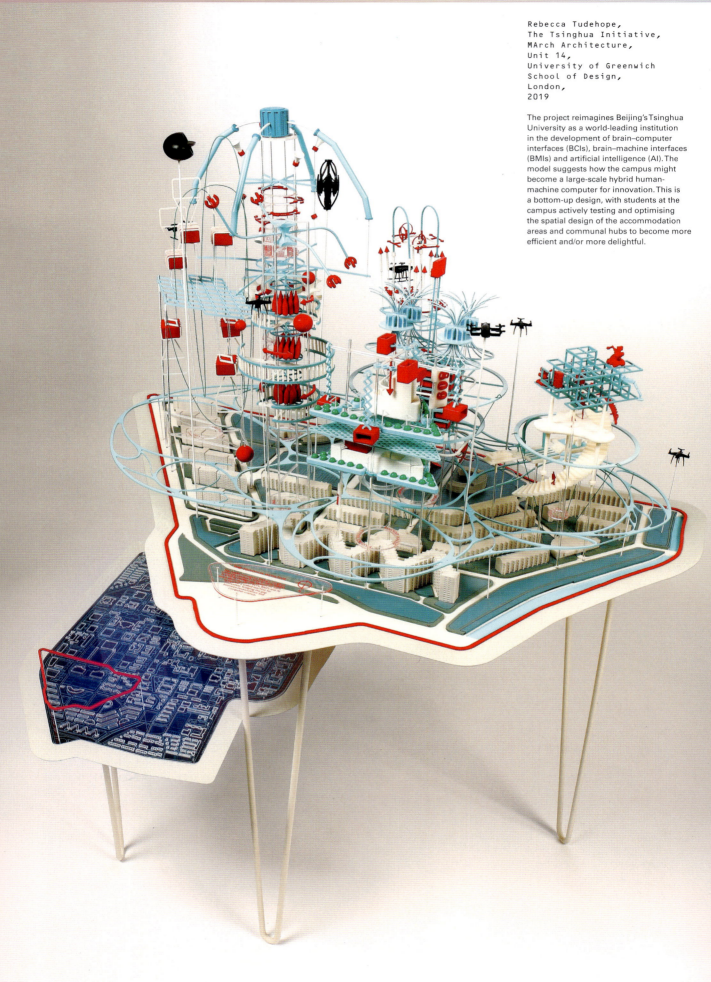

Rebecca Tudehope,
The Tsinghua Initiative,
MArch Architecture,
Unit 14,
University of Greenwich
School of Design,
London,
2019

The project reimagines Beijing's Tsinghua University as a world-leading institution in the development of brain–computer interfaces (BCIs), brain–machine interfaces (BMIs) and artificial intelligence (AI). The model suggests how the campus might become a large-scale hybrid human-machine computer for innovation. This is a bottom-up design, with students at the campus actively testing and optimising the spatial design of the accommodation areas and communal hubs to become more efficient and/or more delightful.

FleaFollyArchitects (pp 48–55) founded their practice with the agenda to explore the potential and limits of narrative-driven architectural models. For them, the model is the architecture, not solely a representational vehicle for a yet-to-be-realised proposal. They take us through a number of their more recent projects that increasingly blur the edges between modelling, installation and architecture.

Historian Chad Randl offers a very different approach to worldmodelling through his research on remodelling (pp 56–63). Every remodel, he contends, is another reality, an alternative world to what was before. Scale has little to do with this qualification of our home as a model of the world; a remodelled interior, a rearranged mantelpiece or shelf. He reminds us that all worldmodels are remodels, as they are all built over and include fragments of preceding ones.

Directing the Architectural Association Design Research Lab (AA DRL) in London, Theodore Spyropoulos considers a series of model investigations that move, sense, transform and amalgamate into architectures, cities, territories and worlds (pp 64–73). These are smart models of potential and agency, each a world unto itself, and each capable of spawning a world.

Following on from their monograph *The Model as Performance* (2018),[7] Thea Brejzek and Lawrence Wallen (pp 74–81) turn their attention to built scenography as both a model of a world and a model for a world. They observe that through deliberately unfinished architectural fragments, the process of worldbuilding has been intentionally interrupted, and these model fragments operate as a self-referential yet autonomous models that provoke discourse between object and viewer. James A Craig and Matt Ozga-Lawn, who co-run the architectural practice Stasus, discuss recent projects that utilise the nature of familiar objects and their deterritorialisation to create complex, performative and imaginative architectures through partial worldbuilding, inferred meaning and mixed-media narratives (pp 82–7).

Walking the line of the prime meridian, Ryan Dillon reveals that the most effective way to model the world is to map it first (pp 88–95). Trudging around Greenwich, London, he finds anomalies and discoveries around what, if anything, should be straightforward. As he reminds us, cartography is creative, and every map a worldmodel. One thinks of the complicated fictive maps of JRR Tolkien or George RR Martin, the latter literally becoming a world model in the famous opening sequence of the television adaptation 'Game of Thrones'.

Phil Ayres discusses his recent EU-funded research into bio-hybrid architectures and argues for the worldbuilding potential of 'coupling' architecture to novel biological systems (pp 96–101). Through this research we see new models emerging of architectures inspired by bacteria, mycelium and insect behaviours. And Art historian Kathy Battista showcases a number of contemporary artists that produce models (pp 102–11). She focuses on how virtual reality is an ideal speculative space, perhaps *the* medium for worldmodelling, that

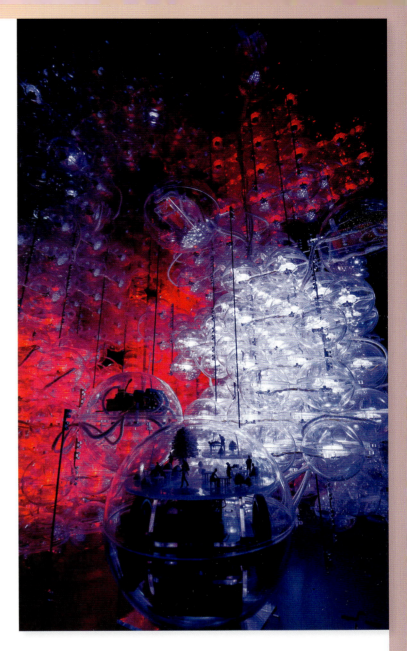

Minimaforms (Theodore and Stephen Spyropoulos), *Emotive City*, 2015

The project imagined a collective and adaptive intelligent system capable of constructing communities based on personal interactions and behaviours. A self-organising system or framework of smart spherical units, each merging infrastructure with inhabitation, could produce whole cities and worlds, but never as fixed assemblies. The extraordinary model reveals both the single architectural cell and one possible amalgamation of many.

opens up a number of architectural possibilities. She notes the accessibility, economy and open-endedness of VR works and, as Hubert also suggests, our increasing willingness to linger in the virtual.

Flipping the Coin
As much as our roles as guest-editors of this ⌭ have been bound up to reaching out to others for their views and expertise on the subject of worldmodelling, much of the joy in this project was had in putting forward our own statements on the topic, for example in examining the history of the British model village (pp 112–19). Originally the invention of architect Charles Paget Wade at the turn of the 20th century, model villages have long fed into our cultural psyche, often bound up in the twee and retrograde. An ongoing model village project for Greenwich, London, is discussed in relation to how this seeming innocence might be misplaced, and how these peculiar enterprises of reactionary architectures, urbanisms and politics might signpost opportunities for how architectural models might function as worldbuilding exercises.

'Paracosmic Project' (pp 120–27) suggests how paracosms seem to answer some basic worldmodelling questions. Who is predisposed to think this way? Why is it useful to nurse a world in the mind's eye over a span of years, a lifetime? How is worldmodelling a genre of imaginative play and a basis for literary, artistic and scientific discovery? The paracosm tips in the psychoanalytic, the blurring of childhood and adult preoccupations, and questions whether such immersive and detailed thinking is not also obsessional and bound to some traumatic experience. The real pleasure found in paracosms is the interplay between fiction writing and crafting fictive worlds, between writing and designing, the storytelling through design as Kate Davies champions. Rather than being framed as a coping mechanism, paracosmic thinking could be considered a workspace in one's mind palace, a synthesis of observations marshalled to interrogate problems and suggest solutions?

There is a firm logic to using worldmodelling to cope and tinker with the world, to expand the scope of architectural endeavour

Mike Aling,
Groenwych for DLR Model Village,
Greenwich,
London,
2020

In the penultimate article in this issue, Mike Aling discusses his own model village project and suggests how eccentric British model villages have perhaps long held many of the clues as to how architectural models might be thought of as worldbuilding exercises.

Wutopia Lab,
Models in Model,
Shanghai,
China,
2019

below: Architecture office Wutopia Lab have recently completed the interior for China's first architectural model museum. As well as being a comprehensive collection of recent Chinese projects in physical model form, the collection is designed as a world of model proposals that together culminate into a vision of a future city named The Last Redoubt.

Worldmodelling

The illustrations in this issue point to the diversity of worldmodels – spanning practice, academia and fine art – as well as the permissiveness of the term. They offer a parallel analysis and suggest how scale is still an important aspect of modelling as an economy of making, but also as a conceptual aid. Even in the paradigm of digital modelling where the subject is assumed to be 1:1, thinking through scaling is still an important, and perhaps necessary, process. If small-scale models in general lend us apprehension more readily and intuitively than, say, plans or sections, the scale of models representing worlds only intensifies this sensibility. We combine two forms of scalar benefits: one suggested by Claude Lévi-Strauss, that 'By being quantitatively diminished, it seems to us qualitatively simplified. More exactly, the quantitative transposition extends and diversifies our power over a homologue of the thing, and by means of it the latter can be grasped, assessed and apprehended at a glance;'[8] the other, more squarely, about things like world models, promised by Gaston Bachelard:

> Such formulas as: being-in-the-world and world-being are too majestic for me and I do not succeed in experiencing them. In fact, I feel more at home in miniature worlds … The cleverer I am at miniaturising the world, the better I possess it. But in doing this, it must be understood that the values become condensed and enriched in miniature. Platonic dialectics of large and small do not suffice for us to become cognizant of the dynamic virtues of miniature thinking. One must go beyond logic in order to experience what is large in what is small.[9]

Yet there is a firm logic to using worldmodelling to cope and tinker with the world, to expand the scope of architectural endeavour, to escape a world in trouble and hopefully return with some ways to rescue it. As Mark Cousins suggests in 'Worlds Without End' (pp 40–47), to work with *the* world, one must fashion for themselves *a* world away. For architecture, arguably models are increasingly proposing less about the object, and more about the objective. ⌂

Notes
1. Albert C Smith, *Architectural Model as Machine: A New View of Models from Antiquity to the Present Day*, Architectural Press (Oxford), 2004.
2. Karen Moon, *Modeling Messages: The Architect and the Model*, Monacelli Press (New York), 2005.
3. Christian Hubert, 'The Ruins of Representation', in Kenneth Frampton and Silvia Kolbowski (eds), *Idea As Model: 22 Architects 1976/1980*, Rizzoli (New York), 1981, pp 17–27.
4. Patrick Healy, *The Model and its Architecture*, 010 Publishers (Rotterdam), 2008.
5. Nelson Goodman, *Ways of Worldmaking*, Hackett Publishing Company (Indianapolis, IN), 1978.
6. Hubert, *op cit*.
7. Thea Brejzek and Lawrence Wallen, *The Model as Performance: Staging Space in Theatre and Architecture*, Bloomsbury (London and New York), 2018.
8. Claude Lévi-Strauss, *The Savage Mind*, Weidenfeld & Nicolson (London), 1966, p 23.
9. Gaston Bachelard, *The Poetics of Space* [*La poétique de l'espace*, 1958], trans Maria Jolas, Beacon Press (Boston, MA), 1994, pp 150, 161.

Text © 2021 John Wiley & Sons Ltd. Images: pp 6–7 © CCA; p 8 © Dennis Maher; p 9(t) © James Lawton. Image courtesy of Mike Aling; p 9(b) © Isobel Eaton. Image courtesy of Mike Aling; p 10 © Isobel Eaton. Image courtesy of Mike Aling; p 11 Photo Theodore Spyropoulos; pp 12–13(t) Image courtesy of Mike Aling; pp 12–13(b) Image courtesy of Wutopia Lab/CreatAR

More on the Model

Christian Hubert

Building on the Ruins of Representation

Studio Klarenbeek & Dros,
Mycelium Chair,
2013

3D-print of living mycelium. Mycelium is the collective product of vegetative (asexual) growth of fungi. It consists of a mass of branching, threadlike hyphae that grow at their tips and respond to environmental stimuli in an exploratory and irregular process. The structure of Klarenbeek & Dros's chair is formed by 3D-printing of the living material, and evokes branching patterns of growth. Actual growth is part of this process as well, as mushrooms will grow out of the mass of mycelium.

New York-based architectural designer and critic Christian Hubert discusses the change in status of the architectural model since he wrote his seminal text 'The Ruins of Representation', published in the *Idea as Model* exhibition catalogue in 1981. The previous notion of the model as object and representation has become much more complex as contemporary concepts of worldmodelling have been investigated and established.

In my desultory manner, I have been preoccupied with models ever since perching at the Institute for Architecture and Urban Studies (IAUS) in New York over the course of the 1980s. The catalogue essay I wrote at the time, 'The Ruins of Representation', along with a brief revisionist postscript written some years later and published in the Dutch journal *Oase*, presented in concentrated form my interest in the expressive registers of architectural models, especially their dual functions as representations and actual objects, in the desires that they embody, and some of their underlying ideological functions in the rhetoric of Postmodernism, with its penchant for historical pastiche.[1]

I have primarily used my training as an architect as a springboard for intellectual inquiry, especially at the IAUS, and for creating environments for artworks, for artists and exhibitions in my design work. More recently I have been working on metal sculptures out of stainless steel. My creative and critical work share implicit affinities that are not always visible, and sometimes they only make sense to me after the fact.

Most models today are more concerned with the future than was the case when I wrote my essay in 1981, and they employ techniques that have developed in the interim – especially digital technologies. One indication of the transformative effects of the digital has been a change in the defining features and ontological status of the model. The model's physicality previously enabled it to function as both object and representation, and to define a space between two and three dimensions. But today that physicality has waned in importance. The physical model has become more of a by-product than it once was. A 3D model produced from a digital file – whether milled, rendered, integrated into the world of a game or even used to fabricate full-sized building parts – remains primarily an expression of its digital instructions. It can convincingly simulate or instruct, but one source of its aesthetic criticality has been undermined. It doesn't really matter if the model is a physical object today. It is the scenario expressed through the model that has become primary, and the goal of modelling a process has taken the place of modelling a product.

Worldmaking
Today, the concept of worldmaking subsumes the idea of the model, which has become an agent of the projective imagination. What worlds do these models build and occupy? Are they a source of hope or fear? What is the role of nature, as human culture has come to consider it, in those worlds? A deep ambivalence informs most of these models. They face the future with a mixture of optimism and dread. They are also meant to directly affect the course of events. Future-oriented models are reflexive, in the sense of creating feedback loops that change perceptions of reality, and in that sense they are meant to change reality itself. They move beyond representation with the goal of setting out possible worlds; some leaning towards forms of utopia, others towards dystopia.

Although the term has not been formally adopted by the International Union of Geological Sciences (IUGS), the current period in Earth's history is informally known as the Anthropocene Epoch, to describe the period when human

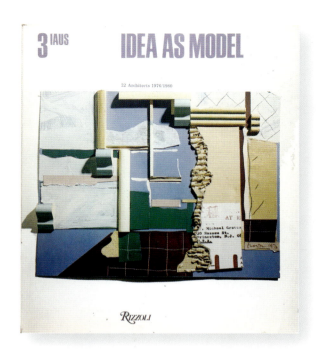

Idea as Model,
Institute for Architecture
and Urban Studies (IAUS)
catalogue cover,
1981

In 1976, the IAUS mounted an exhibition entitled 'Idea as Model'. Paul Goldberger in the *New York Times* would complain of the exhibition: 'What is most disturbing is that there is no attempt, either through choice of exhibitors or through any sort of accompanying text, to discuss the whole question of models and their role in the process of making architecture' (27 December 1976, p 58). The publication of *Idea as Model* (1981) would come after the fact, as a retrospective catalogue, with essays by Richard Pommer and Christian Hubert and a cover collage by Michael Graves.

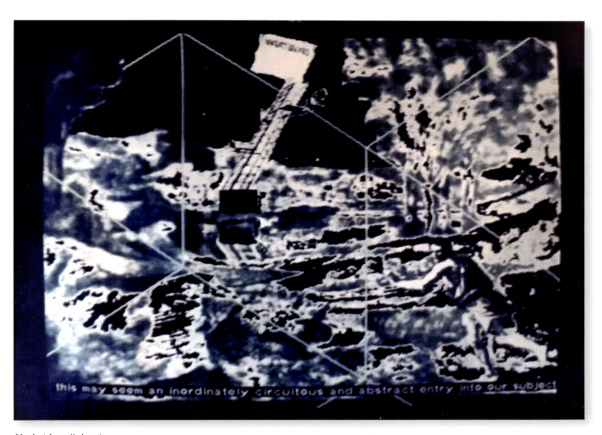

Christian Hubert,
Cuber(t),
an Architectural Folly,
Leo Castelli Gallery,
New York,
1983

This early computer image of a future form of architectural folly evokes an imaginative experience of cyberspace as a historical landscape. The image incorporates a figure from a painting by Nicolas Poussin (*Landscape with Man Killed by a Snake,* 1648), El Lissitzky's *Lenin Podium* project (1924) and a quote from the intellectual historian Hayden White.

Christian Hubert,
Inner Landscape (blue),
Dudley,
Massachusetts,
2019

The play between representation and objecthood has been of explicit interest in painting and sculpture. Following the directions indicated by Picasso's guitar, which translated Cubist syntax into actual space, a number of works have situated themselves at the bounds of pictorialism and objecthood, locating the concerns of art at a crossroads of painting, sculpture and architecture. This painted stainless-steel piece occupies a similar niche.

Christian Hubert,
Transitional Object a,
2020

Paul Goldberger sought to differentiate architects from both painters and sculptors. He claimed that good architects are not sculptors, for architecture is concerned with the creation of interior space, whereas sculpture is pure form, although Frank Gehry's work undermines that distinction. This sculpture, one of a series, suggests a similar ambiguity. Considered as a 'transitional object', a developmental concept developed by Donald Winnicott, it resists answering the question of whether it was found (in the world) or made up instead.

activity started to have a significant impact on the planet's climate and ecosystems. The concept of the Anthropocene gives humans planetary roles and responsibilities. The term is meant to be scientifically descriptive, but it inevitably entails ethical responsibility for its consequences. On the one hand, it acknowledges the defining role of humans: they have become forces of nature. On the other hand, it implicates humans in processes that they do not fully understand and cannot really control. As global 'apex predators', humans themselves are at risk from the environmental stresses they have caused or exacerbated, such as rising sea levels, the fragility of human food-chains, water shortages and the possibilities for global pandemics. In this sense, worldmaking is not a metaphor. It is a literal obligation, and not just to humanity. It requires imagination, but not in the form of wishful thinking.

We have already proclaimed the Death of Nature. Perhaps it is not dead yet (we would be too), but what a job humans have done damaging it. Humans are making a world inhospitable to most species and increasingly to themselves. While they have succeeded in making planet Earth support a greatly increased human population, with significant gains in human standards of living, this achievement seems increasingly fragile. The bills are coming due, and the costs to other species and to habitats will be unbearably high.

The dreams of new worlds seem to coincide with the imminent collapse of the old one, and the new 'desire' of the model, as expressed in worldmaking, is to be instrumental in creating these new worlds, to promote a reflexive reality, informed by hopes and fears that motivate its dynamic force. In this sense, models have become performative. They not only point to the future, but lay claim to building new worlds (even fictional ones), in the full awareness that those will not necessarily be improvements on the existing ones unless humans become stewards of life on Earth.

There are two main ways in which these reconfigurations of reality, technique, aesthetics and natural processes can inform one another. The first is an explicit ecological and social agenda, in which life in every form is understood as process, in which humans, their technologies, other species and the planet are all stakeholders. The second is the project of enabling human inquiry and design to work symbiotically with other agents – other processes or other species. The current task of models is to incorporate those parallel strains, to embody their potentials, and to function as 'models' in the sense of exemplars.[2] A wide range of exemplars is available today. These include hybrid life forms – chimeras combining organisms and human technology, or new subjects such as networked symbiotic organisms (forests, lichens, slime moulds etc) – that move beyond concepts of the individual or population.

Worldbuilding
Like many boys his age, a young man in my immediate family (call him *A*) has spent long hours online playing computer games. It is tempting to call this an addiction, but I feel that this would be ungenerous. It is certainly a

Float of Vladimir Tatlin's
Monument to the Third International,
Leningrad, Russia,
1925

The model purports to present architecture, not represent it. Unlike the signs of language, whose signification is primarily a matter of arbitrary convention, the relation of the model to its referent appears motivated, in the sense that it attempts to emulate or approximate the referent. Here, the model is defined by its performative function, as a harbinger of a new utopian society. This simplified model (of a model) was carried through the streets of Leningrad in 1925. Joseph Stalin may have been on the tribune.

The dreams of new worlds seem to coincide with the imminent collapse of the old one, and the new 'desire' of the model, as expressed in worldmaking, is to be instrumental in creating these new worlds

time sink, though, and he seems to inhabit a parallel world in which he socialises with friends he has never met, even to the point of actually rescuing one of them who had taken steps to kill himself. Over the years *A* has played a series of games, from Minecraft, which definitely involves 'worldbuilding', to Total War. Minecraft is explicitly structured around making worlds, both on the individual and community level, and it is endlessly modifiable. The game is a small step away from architectural modelling programs like SketchUp, and in turn it has modified the way the newer generations perceive architecture. *A* claims, with some truth, that he has learned a lot about military history in Total War, in its many iterations, as the game seems to have been meticulously researched. Games such as this provide vivid experiences of other times and places. In an application to college, he submitted an essay that involved participation in the 19th-century Greek War of Independence, which he was able to describe in convincing terms, due at least in part to his first-person game experiences.

In his extraordinary novel *The Overstory* (2018), which focuses on the life of trees and humans who value and communicate with them, Richard Powers describes a brilliant game designer, crippled in childhood from falling out of a tree. In the book, this genius coder develops a highly successful multiplayer world game called Mastery. His project managers, who have become 'boy millionaires' in the process, think he is crazy when he suggests incorporating 'the marvelous world of what is happening underground, which we are just starting to learn how to see'. They insist you can't make a game out of plants, 'Unless you give them bazookas'. His response is to ask, 'Why give up an endlessly rich place to live in a cartoon map? Imagine: a game with the goal of growing the world, instead of yourself.'[3]

I am reminded of the confusion between the map and the territory, in Jorge Luis Borges's one-paragraph literary forgery 'On Exactitude in Science', written in 1946 and quoted in 'The Ruins of Representation': 'In that Empire, the craft of Cartography attained such perfection that the Map of a Single province covered the space of an entire City, and the Map of the Empire itself an entire Province. In the course of Time, these Extensive maps were found somehow wanting, and so the College of Cartographers evolved a Map of the Empire that was of the same Scale as the Empire and that coincided with it point for point.'[4]

Both examples are doubly fictional – a hypothetical game described in a work of fiction, and a spurious history ascribed to a fictional author – but the 'worldbuilding' impulse remains unmistakable, and they express the same priorities that I have been arguing for here. Even the ambiguities of the 'real' and the 'virtual' recall the arguments of 'The Ruins of Representation'.

Models have become increasingly anticipatory in nature, not as simple representations, but as predictive scenarios. In this sense, they contribute to world 'making'. In the Anthropocene era, they describe the world that humans are in the process of making by extrapolating from current trends in environmental degradation. For the most part, they are adaptations to conditions increasingly hostile to human life, in a hot, dry and dangerous planet, or on other planets altogether.

Christian Hubert,
Kelp Oasis,
2020

Worldmodelling emerges as the world starts to collapse. It is a coping mechanism and a logical reaction at the same time. Only through speculation at the scale of the world can solutions to ecological, climate and socioeconomic catastrophes be considered. This sketch for an 'ocean oasis' proposes a design for an educational kelp farm and wellness centre that offsets the warming and acidification of the ocean at a local level.

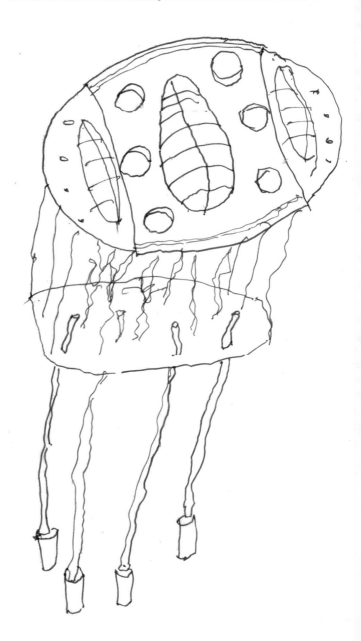

Neri Oxman and Mediated
Matter Group,
Silk Pavilion II fabrication,
Abano Terme,
Italy,
1999

opposite: The Bombyx mori silkworm starts by spinning a scaffolding structure that it triangulates, while attaching its fibres to its immediate environment. Over the course of spinning this scaffolding, it will also close in onto itself to begin to construct its cocoon out of a single fibre. The fabrication process of the Silk Pavilion emulates the process of the silkworm through both robotic techniques and management of live silkworms. It consists of two phases: the creation of the 'scaffold' by a robotic arm, and subsequently deploying thousands of silkworms to spin a secondary silk envelope. A rotating jig ensures that they spin a flat surface. The project authors describe the Silk Pavilion as a case study for biomimetic digital fabrication.

worldmodelling

Nature everywhere speaks to man in a voice that is familiar to his soul.
— Alexander von Humboldt[5]

If worldmaking aims at making and remaking the world, and worldbuilding seeks to develop alternate or imaginary worlds, worldmodelling is a search for insights into the workings of the world as we find it. Apparent triumphs of the technological imagination are taking place against a backdrop of anxiety about species' extinctions and threats to the survival of human life. The title of a sobering account of a time after humans, *The World Without Us* (2007)[6] purports to give a scenario of the pace of change that would result from the end of human intervention in the landscape, and the process of natural self-healing that would take its place. In cities like New York, during the COVID-19 pandemic, we have recently caught a glimpse of that world, when birds are heard singing in the early morning instead of the rumbles and honks of automobiles, or when the sky returns to a bright blue without the haze of pollution.

Even as the natural world is under attack, profound new insights into its workings are suggesting 'models' – in the sense of exemplars – for creative thought and design. The potential for new forms of modelling is not limited to addressing environmental issues. They also open up new modes of aesthetic imagination. Biological evolution, for example, has primarily been described in terms of (functional) adaptation, and judgements of beauty have generally been considered the exclusive province of humans. Aesthetic judgement, especially since the Enlightenment, has been identified with a harmony between the world and the workings of the human mind. But in *The Evolution of Beauty* (2017), Richard Prum convincingly documents the 'transformative power of female mate choice' on the appearance and behaviours of male birds.

For Prum, following Darwin, '"Beauty Happens" (*as the animal perceives it*)' whenever the social opportunity and sensory/cognitive capacity for mate choice has arisen.[7] New opportunities for symbiotic design and co-production with other species are emerging, such as Neri Oxman's and the MIT Mediated Matter Group's work with silkworms (1999), or the mycelium chair by Studio Klarenbeek & Dros (2013), although a delicate creative balance between humans and other species needs to be established in every case. As humans, we need to move away from our species-centric views of nature and our urge to control in order to better understand how we are part of the world.

Alexander von Humboldt claimed that the 'netlike, entangled fabrics' of nature appear gradually to the observer.[8] According to Merlin Sheldrake, what was a metaphor used to describe the 'living whole' of the natural world is literally the case in mycorrhizal (fungal) networks, in which trees and fungi are inextricably entangled. Science and technology have led to new metaphors in turn, such as 'the Wood Wide Web' – coined more or less at the same time as mathematical tools were being developed for the study of networks, and when the World Wide Web seemed to afford new utopian opportunities. In a humorous passage, Sheldrake wonders if humans tend to give ontological priority to trees over fungi, a 'plant-centric'[9] view of fully symbiotic relationships.

A poignant slogan of the counterculture of the late 1960s was 'Utopia Now!' (today a reference to a clothing brand – most likely trademark protected), and it is only a small step from the study of fungi to calls for 'changing one's mind' through psychedelics. A recent book by Michael Pollan, a convert to benefits of psychedelics, documents lasting changes to personality occasioned by consuming 'the flesh of the gods', including the sense of being one with nature, and the feeling of 'co-creatureliness'.[10] This renewal of interest in the benefits of psychedelics has given new hope to wishful thinking, which lies beyond the scope of this brief essay. It remains an open question as to whether subjective experiences of 'one-ness' with nature can lead to effective attunement between nature and culture, but the key question for humanity is whether it can integrate its increasingly powerful capacity to control the Earth with its ethical responsibility for the future of life and a symbiotic and poetic relation to other life forms. Whether the Anthropocene will turn out to be a moment of creativity or a catastrophic one remains very much an open question. Performative models will serve as signposts along the way.

Notes

1. Christian Hubert, 'The Ruins of Representation', in *Idea as Model*, Rizzoli (New York), 1981, pp 16–27, and 'The Ruins of Representation Revisited', *Oase #84: Models Maquettes*, NAi Publishers (Rotterdam), 2011, pp 11–19.
2. See Thomas Kuhn, *The Structure of Scientific Revolutions*, University of Chicago Press (Chicago, IL), 1974.
3. Richard Powers, *The Overstory*, WW Norton (New York), 2018, pp 412–13.
4. Jorge Luis Borges, 'On Exactitude in Science', *A Universal History of Infamy*, Penguin Books (London), 1975, p 325.
5. Quoted in Andrea Wulf, *The Invention of Nature*, Alfred A Knopf (New York), 2015, p 54.
6. Alan Weisman, *The World Without Us*, St Martin's Press (New York), 2007.
7. Richard O Prum, *The Evolution of Beauty*, Doubleday (New York), 2017, pp 72, 119–20.
8. Merlin Sheldrake, *Entangled Life*, Random House (New York), 2020, p 149, and note page 268.
9. *Ibid*, pp 160 ff.
10. See Michael Pollan, *How to Change Your Mind*, Penguin Press (New York), 2018, ch 2.

Text © 2021 John Wiley & Sons Ltd. Images: pp 14–15 © Studio Klarenbeek & Dros / www.dotunusual.com <http://www.dotunusual.com/>; p 17(t) Courtesy of Christian Hubert; pp 17(b), 18, 20 © Christian Hubert; p 21 © Neri Oxman / Mediated Matter Group

Miniature Places for Vicarious Visits

Worldbuilding and Architectural Models

Mark JP Wolf

Frederik and Gerrit Braun,
Miniatur Wunderland,
Hamburg,
Germany,
2011

The Italian section of Miniatur Wunderland, as seen at night, requiring thousands of miniature light bulbs to be integrated into the models and automated by computer. Using lighting to simulate a diurnal cycle adds to the realism of the model as well, and gives viewers a God's-eye view not only spatially but temporally.

Professor Mark JP Wolf from Concordia University Wisconsin in Mequon takes us on a historical journey, discussing significant aspects and precedents of the practice of worldbuilding, from the ancient Egyptians to JRR Tolkien and George Lucas. He demonstrates that worldbuilding is a deep part of the rich history of human culture, and key to humanity's imagination.

Architectural models are as old as worldbuilding itself. The Egyptian collection in New York City's Metropolitan Museum of Art contains objects from the tomb of Royal Chief Steward Meketre (c 1981–1975 BC) which include 24 models – of a bakery, brewery, granary, slaughterhouse, carpenter's workshop and more. The Ancient Egyptians even used architectural models to gain authorisation for building projects, some as tall as three storeys high, and their models included such details as doorways, windows, balconies, stairways, pillars and even large storage jars.[1] Museum exhibits on famous architects, like Frank Lloyd Wright and Frank Gehry, often contain models of buildings as well. And then there is the Archi-Depot Museum in Japan, a museum devoted to the display of architectural models, from study maquettes to final design models.

Apart from their historical and aesthetic value, not to mention their practical value as three-dimensional blueprints for the planning of structures, architectural models also represent a particular form of worldbuilding, a practice most often associated with fantasy and science-fiction creators, like JRR Tolkien or George Lucas. Worldbuilders are more than just storytellers; they often have detailed descriptions or imagery of such things as vehicles, buildings, weapons, clothing and designs present within their works as well (models used for filmmaking, even when they do not appear on screen, are often seen in 'Making Of' books and documentaries, as well as museum exhibits on filmmaking and various franchises). Tolkien has detailed descriptions laying out the design of places like the seven-tiered city of Minas Tirith, whereas filmmakers like Lucas often build elaborate physical models of locations in order to plan film shoots, and many models are actually used in the films as well, as visual effects. Architectural models are a unique area of worldbuilding because, like the models of fictional worlds, they begin as imaginary places, designs being constructed in three dimensions, but instead of remaining imaginary they bring those designs into the real world, and the places they depict become real places. Imaginary worlds have always overlapped the real world (or 'Primary World', as Tolkien called it); in many stories travellers pass from our world to an imaginary one, either by vehicular transportation, portals or other means connecting the two worlds in some way.[2] Although they are objects in the real world, architectural models are likewise depictions of imaginary places, ones that often precede the real buildings which are made in their image. Thus, the architectural model is both a form of worldbuilding as well as a tool for worldbuilding, an artwork as well as an intermediate stage leading to a final construction. But architectural models can also be built based on existing buildings, or built just for their own sake, without ever leading to a full-scale building.

Models as Miniature Worlds
As the Met's Egyptian collection demonstrates, architectural models have a long history. Their entry into more public life came in the 1700s and 1800s, when dolls' houses were produced not as toys but as educational tools used by mothers to teach their daughters household management.[3] Some later dolls' houses were built as miniature replicas of the owner's house to show off their wealth. By the end of the 19th century,

Model of granary with scribes from the tomb of the Ancient Egyptian Royal Chief Steward Meketre, Metropolitan Museum of Art, New York, c 1981–1975 BC

The granary model is populated with characters demonstrating how the areas of the building are used. It shows the bags of grain being recorded by scribes; another man carries a bag through the door from the scribe's room to the storage room (with several others in line behind him); and in the storage room five men are emptying their bags into the room-sized storage areas.

Model of a slaughterhouse from the tomb of the Ancient Egyptian Royal Chief Steward Meketre, Metropolitan Museum of Art, New York, c 1981–1975 BC

Although they have doors and windows to the outside, these Egyptian models are handily contained within the wooden boxes that serve as their outer walls. Not pictured here, the slaughterhouse also had a removable roof, a wooden lid that covered the top of the model.

Frederik and Gerrit Braun,
Miniatur Wunderland,
Hamburg, Germany,
2011

In Miniatur Wunderland's 150-square-metre (1,600-square-foot) Knuffingen International Airport section, details on rooftops and inside windows add to the model's realism even if one does not consciously note all the effort that has gone into them. Also, unlike model railways which are mass-produced and sold commercially, many of the Wunderland models were custom works that had to be built by hand.

Narcissa Niblack Thorne,
English Drawing Room
of the Early Georgian Period,
Thorne Miniature Rooms,
Art Institute of Chicago,
c 1937

An example of one of the highly detailed rooms in the Thorne Miniature Rooms. Besides the elaborate miniatures that make up the décor, the rooms are lit to match the lighting of full-sized rooms as they would appear at various times of day, making the illusion of scale even more successful.

when industrialisation allowed objects to be mass-produced, dolls' houses came to be considered objects for play, with miniature people used as avatars, allowing children to vicariously enter the world and interact with it.[4] At the same time, dolls' houses and miniature buildings would also remain an art form, as in the collection of painstakingly detailed models of the Thorne Miniature Rooms collection, created in the 1930s and on permanent display at the Art Institute of Chicago.

Contemporaneous with dolls' houses was another hobby involving miniatures, which is also sometimes elevated to an art form: model railways. The first mass-market model train sets, made by Märklin in Germany, appeared in 1891; trains running on a metal track came in 1896; Lionel's first electrically powered train came in 1901; and by the 1920s, model railways had become a popular hobby.[5] While trains are the central element, entire landscapes with buildings, even cities, are often a part of such models. Today, elaborate train sets can be found in museums, like *The Great Train Story*, a 325-square-metre (3,500-square-foot) model railway on permanent display at Chicago's Museum of Science and Industry since 2002, or the many models of the London Transport Museum in England. In Hamburg, Germany, the Braun brothers' Miniatur Wunderland tourist attraction is a model world installed in 2011 throughout a three-storey warehouse. These elaborate, detailed landscapes include a model railway with 930 trains running on more than 13 kilometres (8 miles) of track, with 215,000 model people, 250 computer-controlled vehicles, and the 150-square-metre (1,600-square-foot) Knuffingen International Airport.[6] Not only are the scenes full of animated movement, but the time-of-day cycles through day and night every 15 minutes, employing 335,000 LED lights installed in vehicles and buildings.[7] Like dolls' houses, these worlds are populated ones, and their animation lends them a temporal dimension that makes their worlds more immersive than static models.

During the rise of dolls' houses and train sets, playsets began appearing, presenting miniature buildings and accessories for children. Begun in 1919, the Louis Marx & Co toy company made metal playsets during the 1930s and 1940s, like the Sunnyside Service Station (1934) and the Roadside Service Station (1935). These were made of metal until the development of plastics made production easier and cheaper, and playsets rose in popularity accordingly. A variety of playsets would continue to be made up to the present day, but a turning point in their history was the introduction of LEGO® building sets. LEGO combined the building set (like Tinkertoys or Lincoln Logs) with playsets (which usually represented one particular setting only), to allow children to build not only the images seen on the set's box, but also whatever they wanted to build out of the bricks included. Because all the pieces could be connected with each other, multiple sets could be combined for even greater building possibilities. Not surprisingly, the LEGO Group has produced hundreds of LEGO models of existing buildings, for their LEGOLAND themes parks and centres as well as for travelling exhibitions for shopping malls and other venues. With a wide variety of pieces, LEGO can be used by children or adults for the building of architectural models, and LEGO even has its 'Architecture' sub-brand,

Like dolls' houses, these worlds are populated ones, and their animation lends them a temporal dimension that makes them more immersive than static models

which consists of dozens of sets based on iconic buildings of the world.

During the 1970s and 1980s, computer graphics became another tool for the building of architectural models, and for worldbuilding as well. They also added something exciting to the models: the ability to visually walk or fly through them, providing a first-person perspective of someone experiencing the models from within, instead of just producing something viewed from outside at a distance. (Eventually, when cameras became small enough, this could be done with model railways as well.[8]) This kind of vicarious experience is evident even in some of the earliest computer-generated architecture, which can be traced back as far as Peter Kamnitzer's *City-Scape* (1968), a 10-minute colour film made at the Guidance and Control Division of NASA's Manned Spacecraft Center. The film simulates streets of block-like buildings to produce a cityscape which is seen and travelled through at eye level. It begins with a simulated moving camera shot flying towards the city's skyline, and soon the point-of-view enters the city, moving around buildings and along freeways, inside of buildings, and even up and down in a glass elevator. Describing the film, Gene Youngblood enthusiastically writes:

> Only a few minutes have passed before a strong sense of location and environment is created … One actually feels 'surrounded' by this city, though viewing it through a porthole. The true three-point perspective invests the image with a sense of actuality even stronger than in some conventional live-action films.[9]

Architectural simulations have grown to become an indispensable tool not just for designing buildings in three dimensions, but also for testing stresses and other physical properties that affect the feasibility of a structure. Thus, such models are also interactive, in ways that even more closely model the actual buildings being designed.

Both physical models, and virtual models to an even greater degree, can simulate other things as well, such as

ow buildings will react to light and shadow, both inside and outside, and even how they might be affected by weather, such as how rain will run off them or how they will be impacted by winds, or how they might look in a surrounding landscape with seasonal changes. While virtual models are now able to model things in far greater detail than a physical model could, they have, at the same time, arguably created a greater appreciation for physical models and the enormous amount of work that can go into them. Thus, during the decades when virtual modelling was developing, we can see a rise in the number of architectural models on display in museums and other places. What is more, virtual architectural modelling has now spread beyond architectural firms, and can be done by anyone with computer animation software, from programs designed specifically for home and interior modeling.[10] Even children have become virtual architects, thanks to video games like Minecraft (2009).

Unlike the worldbuilding activities involved in the building of imaginary worlds, architectural modelling has often taken designs from the imagination of designers and made them actual structures in the real world. But ever since the days of commercial dolls' houses, and today more than ever, such modelling, whether physical or virtual, is not done with the goal of a full-size building in mind, but rather just for fun. And that can be attributed partly to the appeal of building miniature worlds.

The Appeal of Miniature Worlds

Miniature models can be found on display in museums and elsewhere: in Disney's Hollywood Studios, there is a display of upcoming lands and attractions at other parks; while the City Hall of Hannover, Germany, has elaborate scale models of the city as it appeared in 1689, 1939, 1945 and 2014. Like most detailed models, or dolls' houses or train sets, one stands gazing at an immense and immersive amount of detail spread out through one's field of view. The viewer of such layouts is naturally given an aerial, God's-eye view of the miniature landscapes in question, which are often designed to

DASSAULT Systèmes,
Virtual room created in HomeByMe,
2020

An image of a room created with the free software program HomeByMe, launched in 2011. To be successful, such programs need not only to simulate the space and the furnishings, but also the lighting to generate a feeling or mood. With enough detail, such images look almost photographic, lacking only the slight imperfections and irregularities found in even the cleanest and most perfect physical rooms.

be impressive in both the size and the amount of detail that they contain. This change of scale can be enjoyable because it makes the viewer a giant by comparison, looming effortlessly over the miniature worlds, which can still be room-sized or larger despite their small scale. This sense of visual mastery, taking in the whole landscape in a single view, yet from a position by which the smallest of details can be appreciated, is part of the pleasure, as is the kind of spectacle that only a physical model can achieve.

Digital, virtual models, on the other hand, allow one to vicariously enter and travel through them, seeing everything from a first-person point of view. If the imagery can be generated in real time, such models can be explored interactively, creating an even stronger illusion of an actual place. Video games often are the best examples of such spaces, presenting cities and different types of terrain in changing lighting and weather conditions, to a scale and degree of detail surpassing physical models. Triple-A games like Grand Theft Auto V (2013), Marvel's Spider-Man (2018) and Red Dead Redemption 2 (2018) feature vast, detailed landscapes that players will probably never see in their entirety. Marvel's Spider-Man, for example, recreates New York City's iconic buildings and boroughs, down to store window displays and garbage in the alleyways and streets.

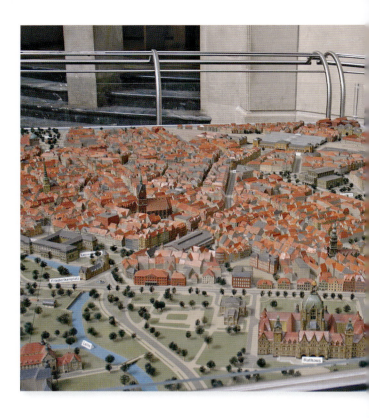

Insomniac Games,
Marvel's Spider-Man,
2018

A screenshot from the re-creation of New York City in the game Marvel's Spider-Man, demonstrating the great visual depth of the scenery. Earlier programs that lacked the computational power to depict a large city stretching out to the horizon tended to limit such vistas, either by blocking points of view or using haze or darkness to limit how far into the distance objects could be seen.

Model of the city of Hannover in 1939, Hannover City Hall, Germany, 2012

A section of one of the city models of Hannover on display in the City Hall there. The four different models, representing different years in Hannover's existence (1689, 1939, 1945 and 2014), together form an engaging narrative history by showing the growth, destruction and rebuilding of the city.

Both physical and virtual miniature worlds gain their appreciation through detail and scope – an enjoyment of the many tiny, carefully crafted parts that together comprise an encompassing whole which binds them all together, our minds filling in the unseen details once a high level of detail has been established: what I have referred to elsewhere as a *world gestalten*.[11] Of course, full-scale buildings can also function to inspire awe and a sense of historical depth, acting as they do as repositories of culture and monuments to other eras.

In an article in *Architectural Review*, Peter Buchanan discusses the psychological necessity of architecture, and architecture as a means of creating ourselves, writing:

> by projecting our psyches into space ... we not only create ourselves but also surroundings to which we sense a strong relationship so we feel at home ... This is taken to an extreme in some sacred precincts or structures that are shaped as a microcosm, a miniaturisation of the cosmos.[12]

Buchanan is referring to architecture of full-scale buildings, but even here, he notes that they themselves can sometimes represent a 'miniaturisation of the cosmos'; thus, architectural models can take miniaturisation even further, resulting in compact representations that condense more than simply buildings, or towns, or cities, but contain entire worlds. And that is the nature of worldbuilding, after all, whether in novels, movies, video games or architectural models: detailed vistas balancing mystery and expectations, scope and detail, giving us enough information so that we can fill in everything else through speculation. The result is an imaginative experience that plays with scale, offering a broad overview and at the same time an intimate view, amazing us with its aspirations to realism even while it blatantly revels in the masterful artifice and meticulous craftsmanship which remains on conspicuous display. ∞

Notes
1. See Galal Ali Hassaan, 'Mechanical Engineering in Ancient Egypt, Part XXVI: Models Industry (Cattles, Butchers, Offering Bearers, and Houses)', *World Journal of Engineering Research and Technology*, 2 (5), September 2016: https://www.wjert.org/home/archive_show/2016/16/VOLUME-2-SEPTEMBER-ISSUE-5.
2. See Mark JP Wolf, *Building Imaginary Worlds: The Theory and History of Subcreation*, Routledge (New York and London), 2012, pp 62–4.
3. See Nicole Cooley, 'Dollhouses Weren't Invented for Play', *The Atlantic*, 22 July 2016: https://www.theatlantic.com/technology/archive/2016/07/dollhouses-werent-invented-for-play/492581/.
4. See Wolf, *op cit*, pp 138–45.
5. See 'A Historic Timeline of Model Train Kits', *ExactRail*: https://exactrail.com/pages/a-historical-timeline-of-model-train-kits.
6. See Rob Lammle, '5 Model Train Sets That Won't Fit Under the Christmas Tree', *MentalFloss.com*, 18 December 2012: https://www.mentalfloss.com/article/31942/5-model-train-sets-won%E2%80%99t-fit-under-christmas-tree. On Miniatur Wunderland, see the official video at https://www.youtube.com/watch?v=ACkmg3Y64_s.
7. *Ibid*.
8. For an example, see Museum of Science and Industry, Chicago, 'The Great Train Story: 10th Anniversary Ride', 23 November 2012: https://www.youtube.com/watch?v=I9vWvPMJudM.
9. Gene Youngblood, *Expanded Cinema*, EP Dutton & Company (New York), 1970, p 254.
10. See '6 of the Best Free Home and Interior Design Tools, Apps and Software', *HouseBeautiful.com*, 28 July 2019: https://www.housebeautiful.com/uk/renovate/design/a28461218/best-free-home-interior-design-tools-apps-software/.
11. See Wolf, *op cit*, pp 51–60.
12. Peter Buchanan, 'The Big Rethink Part 4: The Purposes of Architecture', *The Architectural Review*, 27 March 2012: https://www.architectural-review.com/essays/campaigns/the-big-rethink/the-big-rethink-part-4-the-purposes-of-architecture/8628284.article.

Text © 2021 John Wiley & Sons Ltd. Images: pp 22–3, 26(t) Photos courtesy of Miniatur Wunderland; pp 25, 26(b) CC0 1.0 Universal (CC0 1.0); pp 28–9 © DASSAULT Systèmes; pp 30–31(t) Photo courtesy of Cara Putman and available at http://www.caraputman.com/dreams/effects-wwii-hannover-germany/

POLYPHONIC

STORYTIME IN SYNTHETIC

Unknown Fields,
The Breast Milk of the Volcano,
film still,
Bolivia,
2018

This is the material of our seemingly immaterial electric future. The world's largest salt flat, the Salar De Uyuni, home to the largest untapped reserve of lithium, a key ingredient in lithium-ion batteries. This is the feeding ground of our green-energy future. According to indigenous myth, this white expanse was created from the mixing of the tears and breast milk of the mother volcano, Tunupa. We power our future with the breast milk of volcanoes.

Kate Davies

DREAMS

REALITY

Reminding us of the importance of the narratives and stories we tell each other and their numerous effects on us, **Kate Davies,** co-founder of the Unknown Fields travelling design studio, and art and architecture collective Liquid Factory, introduces us to architecture that is a synthesis of the real and virtual worlds that are starting to seamlessly permeate our environment.

Unknown Fields,
In Red Ochre,
film still,
Australia,
2019

Worm's-eye Lidar scan animation showing an open-pit mine from below. *In Red Ochre* is a film about 'knowing' the Earth. Aboriginal narratives speak of the Dreamtime when the ground was soft and creation beings shaped mountains and rivers, and Aboriginal paintings reveal the deep mythological significance of the land to the people who hold it sacred. Meanwhile, many of our future technologies will contain minerals still (for now) embedded in the same red soil of the Australian outback. A new wave of mineral exploration is being driven by the demand for raw materials that are essential to emerging technologies.

Stories, like music, are dangerous. They're contagious. They get passed along, pressed into your hands and then into your head. The really good ones infect you for decades afterwards.

— Lauren Beukes, 2014[1]

We are made of stories, they are the texture of it all, they build our models of the world. The stories we tell ourselves shape our identities, collectively and individually. Stories form a structural logic for how we act in the world, a useful connective tissue with which to make sense of reality. They unfold complexity. Stories embroider plain old chaos with the seductive beauty of cause and effect, operating artfully in the spaces between things, as a kind of cultural infrastructure, stitching significance into the landscape of the everyday.

In this way, stories act as maps of systems. They are tools for systems thinking, which makes them a critical site of operation. In order to address the deep issues we face, we need to make systemic change. To do that we need to think systemically, which means we really need to see systemically.

Stories are also a mechanism for transcending the limits of human perception, extending our experience beyond the here and now into the past or the future, to places unvisited and people unknown. They connect the dots and fill in blanks – for better or worse. Stories encode value systems. Stories are addictive and infectious. Stories are dangerous. They influence our identities and they reinforce biases.

Unknown Fields uses expedition-based research to construct a series of narrative journeys into the shadows cast by the city. The practice makes films and physical work that reveal a complex global network of landscapes that are set in motion by the city's desires. Not only the physical spaces, but the multilayered virtual realms and systems that enmesh them (data landscapes, digital environments, myth, poetry and territories of the imagination).

Stories are an operating system for culture. We need storytellers, to build their models of the world, to illuminate what is there and to prototype what is not. We need voices to sing a better world into being. Like the musical texture of polyphony, stories are a cultural texture. Many stories by many voices, unfolding over time.

More than 59 zettabytes (ZB) of data was created, copied and consumed globally in 2020.[2] This 'Global DataSphere' is the raw material of humanity's story repository. The construction of reality is a project of collective authorship, narrated en masse through a vast planetary media network. Ever-more people sharing ever-more stories, reaching ever farther and ever faster. These are fragmented worlds, reflected in a mirror ball of simulation, manipulation and reappropriation. This highbrow/lowbrow recycled soup of visual junk, swilling around the Internet, coagulating into a temporary collective narrative, is the confused memory of a drunk culture, a gigantic oral tradition.

Songlines in the Server

The planet is wired, cocooned in a vast sensor network of satellites, drones, microsensors and smartphones that construct what Jennifer Gabrys calls 'Program Earth'.[3] The ultimate model of the world is out there, written in the data; we just cannot see it properly – yet.

Big data means increasingly complex datasets generated at an accelerating pace. Data science emerged to extract patterns – 'narrating' the data. Machine learning brings exponentially faster analysis, extending our human perception to include global senses, with astonishingly powerful tools for distilling meaning from the noise.

Big data, virtual reality and artificial intelligence are converging in a nuclear-strength explosion of the habitable world model – the 'digital twin'

Unknown Fields and Toby Smith,
Rare Earthenware,
film still,
China,
2015

above: The film charts a reverse journey, unmaking our electronic devices, from the high street, back across oceans aboard a cargo ship, through China's vast wholesale market in Yiwu, and manufacturing centre in Shenzhen, and on to the refineries and mines in Inner Mongolia that bring us rare earth minerals – critical ingredients in all those luminous devices in our pockets.

> But the model is not the world – to say the model is habitable does not mean we should go and live there. It is just a powerful way of seeing the world

Unknown Fields,
View from the bridge of the *Gunhilde Maersk* on location for *Rare Earthenware*,
Yantian Port,
China, 2015

below: Oh, lovely thing, how far have you voyaged to be with me? The sea has become the bearer of things. Though they seem to arrive by magic in smiling boxes, somewhat surprisingly, 90 per cent of the objects that surround us have travelled by sea. All these mundane bits and bobs, once brave seafarers, hauled in heaving ships on extraordinary voyages – an absurd narrative of ordinary things.

We are jacked into the 'planetary sensorium',[4] this giant observatory of tins of beans and blue jeans, hurricanes and riots. Perhaps we should be marking this as a perceptual shift of extraordinary importance – a sense extension, sense substitution. If the blind can learn to see with their tongues – via tiny video cameras connected to pulsing postage-stamp-size lollipops in the mouth – given time for the brain to interpret the inputs, will we be able to feel the planet? Or perhaps it will present itself as some sort of data synaesthesia.

One of the defining projects of Big Data is the drive to produce a Digital Earth,[5] a virtual globe constructed of massive, multiresolution, multitemporal, multi-layered Earth observation data combined with relevant analysis algorithms and models that can perform reconfigurable system simulations ... for complex geoscience processes and socioeconomic phenomena'.[6]

So, all our models will (very imminently) have the potential to be plugged into the most sophisticated model of all time, in real time. Skinny-dipping in the flow, with eyes wide open to contingencies and relationships. With enough data, machine learning does the grunt work, effortlessly generating astonishingly detailed simulations with the potential for multiple 'save-as' parallel realities – infinite speculative divergences. The simulated multiverse.

So, what does this God's-eye view enable? And what happens when we let the poets, the thinkers and the dreamers loose inside the machine?

Storm Chasers + Synthetic States
Couple the power of that all-seeing data stream with the high-definition eye-popping pixel-fest that we all bathe in and we are entering a lucid dream state. But the model is not the world – to say the model is habitable does not mean we should go and live there. It is just a powerful way of seeing the world. This may be easy to forget when the model is a fully immersive world-within-a-world, when its 'reality' gets inside your head, when it is the *Vurt* feather, when it is more drug than image. When the siren servers[7] lure you onto the rocks of simulation and down, down into the deep, deep fake.

Big data, virtual reality and artificial intelligence are converging in a nuclear-strength explosion of the habitable world model – the 'digital twin'. A whole-world simulation, inhabitable at multiple scales and timeframes with endless parameters for running scenario testing.

In September 2020, Microsoft released its new flight simulator, a game in which a full reconstruction of planet Earth exists – topography, cityscapes and weather systems – in real time. It is driven by live satellite imagery, geo-sensor networks and meteorological modelling. If a volcano erupts, you can fly over and take a look. This is a game, remember. It had been out three days and the storm chasers were already out in force. Catastrophic Hurricane Laura was making landfall in Texas and Louisiana, modelled in real time through sophisticated simulations drawn from the data, with people flying into the live eye of a devastating storm. Perhaps we all are.[8]

Purveyors of Fine Data and Deep Fakery
Somewhere between this stitching of reality into a game, and the reverse, outrageous deep fakery – an AI-generated deep-fake video of Barack Obama saying something scandalous for example – is the emerging world-model explosion, a synthetic

50° 43' 38.3" N
3° 28' 28.5" W

Human Conveyor Belt
Maintenance Cost:
€100 /day

Mechanical Machine
Maintenance Cost:
€3800 /day

Unknown Fields,
Snowing in the Supercomputer,
from *The Dark Side of the City* book series,
2016

above: The Cray XC40, the UK Met Office's supercomputer and one of the fastest in the world, spends its time crunching climate models. This computer is in Exeter, but it contains the world. It spends a great deal of its time thinking about the Arctic. Out there on the pack-ice, under the midnight sun, is a landscape that is home to a people who are observing its changes unfolding in real time. In this white room, there is its digital twin. Nobody lives here.

Unknown Fields,
All Up in My Grill, film still,
Ilakaka,
Madagascar,
2016

opposite: In the Wild West mining town of Ilakaka is a human conveyor belt, bodies moving in rhythm to extract precious gemstones from dry dirt. These stones, pulled by the many, end up adorning the necks of the few, flashed in audacious displays of wealth, on yachts, at parties and on camera – in the jewelled gold teeth of superstar bling. The men who extract them are paid in rice. As the gems pass from hand to hand, their value is transformed. A web of exchanges crosses the planet, twisted and absurd.

cut-and-shut. In this hybrid reality, the edges of our models are dissolving, holes of reality opened up inside them. And conversely back in real life, synthetic imposters, deep fakes, algorithms, neural networks and bots are rewriting scripts, re-authoring the world.

The high-fidelity cheat-feats of technical artistry that AI ushers in, where every representation of the world can be effortlessly and immediately perfect – indistinguishable from real life – has well and truly cut the moorings on reality, if there ever was such a thing. This is total presence,[9] mainlining synthetic states of body and mind straight into the prefrontal cortex.

Our models feed off the material world, and they can chew it up and spit it out. For the neural networks of AI to work, you have to spoon in datasets, images and videos. They learn from how the world is, not how it could be. AI is not a dreamer, it is a vampire. We feed our machines our own values. Evidence of racial and gender bias has been found in machine-learning algorithms.[10] The cold logics that underwrite everything we know are infused with implicit biases and unconscious values that have disturbing implications.

Who is writing the script? The more data you feed AI, the more accurate it becomes. The search for good-quality AI 'training data', verified for authenticity and accuracy, has become a new sport. The best data is rich in detail. Like a fine wine, fine data – a good vintage. Our AI overlords deserve the best.

But the thing about data-driven models is that they cannot grasp the most intangible parts of human experience. There will always be a resolution gap. We should not forget to hold the model to account for the things it cannot easily portray. The model is a tool, not a destination. The data is not the story; it is just the raw material for the story. And it might be a fine vintage, but it really needs to be drunk by the poets and dreamers, the radicals and transgressives. Because the true model of the world is not generated by algorithms; it needs to be a glorious polyphony, sung forcefully into being. Stories hold things that nothing else can measure. ᗅ

Notes
1. Lauren Beukes, 'Introduction', in Jeff Noon, *Vurt* [1993], Tor (London), 2014, p 7.
2. 'IDC's Global DataSphere Forecast Shows Continued Steady Growth in the Creation and Consumption of Data', IDC, 8 May 2020: www.idc.com/getdoc.jsp?containerId=prUS46286020.
3. Jennifer Gabrys, *Program Earth: Environmental Sensing Technology and the Making of a Computational Planet*, University of Minnesota Press (Minneapolis, MN), 2016.
4. Lukáš Likavčan and Digital Earth, 'Searching The Planetary In Every Grain Of Sand: Introduction To Digital Earth Fellowship 2020–2021', *Medium*, 1 June 2020: https://medium.com/digital-earth/searching-the-planetary-in-every-grain-of-sand-introduction-to-digital-earth-fellowship-2020-2021-8692e5ff3a05.
5. The concept of Digital Earth was first coined in Al Gore, *Earth in the Balance*, Earthscan (London), 1992.
6. Zhen Liu *et al*, 'Understanding Digital Earth', in Huadong Guo, Michael F Goodchild and Alessandro Annoni (eds), *Manual of Digital Earth*, Springer (Singapore), 2020, p 3.
7. Jaron Lanier, *Who Owns The Future?*, Penguin Books (London), 2014, p 55.
8. Tom Warren, 'Microsoft Flight Simulator Players Are Flying Into Hurricane Laura', *The Verge*, 27 August 2020: www.theverge.com/2020/8/27/21403769/hurricane-laura-microsoft-flight-simulator.
9. Lutz Jäncke, Marcus Cheetham and Thomas Baumgartner, 'Virtual Reality and the Role of the Prefrontal Cortex in Adults and Children', *Frontiers in Neuroscience*, 3 (1), 2009, p 52: www.frontiersin.org/article/10.3389/neuro.01.006.2009
10. Martin Ford, *Architects of Intelligence*, Packt Publishing (Birmingham), 2018, p 2.

Text © 2021 John Wiley & Sons Ltd. Images: pp 32–5, 38–9 © Unknown Fields; p 36(t) © Unknown Field/ Toby Smith; pp 36–7(b) © Photo Kate Davies

Ayham Jabr,
Retro Landscape,
2019

A world model may be inhospitable, broken, a world in need of repair; thus a world set up as a problem. Science-fiction author Douglas Adams described a world, Magrathea, where hyperspatial engineers and designers fashion custom-made luxury planets for impossibly wealthy clients. Magrathea itself falls into ruin after bankrupting the rest of the universe.

Mark Cousins

worlds without end

In this, his last essay, the late iconoclast **Mark Cousins**, former Head of History and Theory Studies at the Architectural Association in London, explores the gaps between making architecture, abstracting it in model form, talking about it and writing about it. He reveals a schism in which words become slippery as we inhabit multiple worlds.

Words sit uncomfortably in the practice of architecture. There must be a better way to describe it – drawing the design, modelling the design, designing the design. They all stress the importance of design as a testimony to the intellectual character of architecture. But they also reflect the fact that as a topic there is no agreed language with which to describe architecture, even in comparison with music, poetry or painting. Every attempt to describe a building is accompanied by the feeling that this is the first time that anyone has tried to do it and that there seems to be no authorised format of description. This surely is the cause for a certain dimension of unhappy life, most discernible in students who tend to be more optimistic. The presentation of their work to teachers and colleagues is often surprisingly halting and unsure. Not so much unsure of what they are presenting but of the whole act of presenting it. As if the task of presentation had arrived with no model of what to do. No one wants to be told what to say, but everyone feels they would like a model of what, or rather how, to say it.

This anxiety about the relation between language and someone's own design has a number of consequences which become obstacles to the student's progress. The first is to dash to the model which becomes the repository for all the energy floating about as anxiety. As a result, the model becomes a detailed, measured, obsessional object. We are close to the world of model trains, aeroplanes, ships in bottles. The model becomes the neurotic object bristling with demands and accuracies, the pointless perfection which fits everything except what is wanted. The second is to leave a permanent sense of lacking a model in the work of an architect – a feeling that there is something unauthorised about their work – not that it is stupid, but that it is a mess. And we should acknowledge that many of the ways in which models have been used, especially in architectural education, have not been successful in developing a happy talent for design. But now the model should be free of so many previous constraints that it should again become a friend of the student.

Ayham Jabr,
Finding the Moon,
2016

A world is composed of elements from other worlds. The collages of Syrian artist Ayham Jabr play this theme out in several directions. The assembly of disparate partial images generates new wholes, new worlds. Photomontage and digital collage combine to compose works that are often illustrative of worldmaking. The method to make a picture and its pictorial content are aligned. Fragmentation and recollection are central to Jabr's artistic practice and spring from his reading of his birthplace and home, Damascus.

The model becomes the neurotic object bristling with demands and accuracies, the pointless perfection which fits everything except what is wanted

We have to recognise that only on the condition that we abandon the concept of the world can we possess the concept of a world

yham Jabr,
ndie Landscape,
017

br, a devotee of science-fiction films since boyhood, considers 'an outer space world' the site of his imaginative practice. The premise of so many sci-fi movies, the colliding outer space and terrestrial worlds is central to much of his work. The inevitable conflict such narratives, worlds colliding, permits analysis and commentary on the conflicts at have ravaged Syria at a remove or critical distance.

Yet there is an enemy of design which must be vanquished before the student can regain a relaxed entry to design. It appears first as a word, but its use has enormous ramifications for how students approach design. It is repeated by almost all students and also by teachers. But the word which turns out to be indispensable is also the word you would most like to get rid of. My own candidate is the word 'idea'. Was there ever a word so vague, so idle, so indifferent to the work of a word as this? Think of its appearance in intellectual discussion; it seems to pop up everywhere. And what is it? Something like a concept (only it is not). Having one (an idea) is like a psychological event and a self-contained statement at the same time. The longer you stare at the word, the emptier it gets.

Anything that dents this hole gives us hope. If I see that someone refers to the 'architecture' of thought or even the organisation of thought, I have a brief rush of optimism. But these words rarely pay their debt. We usually find ourselves shunted into the world of metaphors, those bankrupts in the world of argument. Only rarely does someone take the path of 'organisation', 'arrangement' or 'architecture' and mean it. Begin to design what had been the antithesis of design? What do we call this – the idea? – this thing which has nuts and bolts, and colours, safety pins, distinctions, short-cuts, musical quotations? The American philosopher Nelson Goodman calls it a 'World' or rather, 'Worlds'.[1]

The End of the World
What in the world is a world? What is the world? Where is it? We start the question by having to assign an unusual importance to the pronoun – *the* world. It matters very much whether we speak of the world or a world. We have to recognise that only on the condition that we abandon the concept of the world can we possess the concept of a world. The world presents itself as the one and only world, the thing that we make scientific statements about, the site of what we call reality, and one which cannot brook any competition.

Each world is made from things (in a traditional sense) and from connections that draw them together

Ayham Jabr,
Surah,
2018

What sparks the creation of a world? A desire to escape from the world? Introspection leads to projection of so many possible worlds. A landscape deftly presents a world owing to habits in viewing landscape art and the economy of a framed scene. Genres of traditional Chinese landscape brush paintings split between those that depicted reality painted by painters versus fantasy landscapes created by the *literati*.

Reality comes with the warning that no other version will be tolerated. At the same time, we imagine that the world should come in a more plural way, many worlds. That does not mean that we inhabit a particular world and live in it in isolation. We live in many worlds and worlds are constituted from what is already around and from altering what is there. Reality is the hum of construction as worlds are made, destroyed, converted and renovated. Each world is made from things (in a traditional sense) and from connections that draw them together. In describing construction we do not concentrate on the objects in the construction itself but on their relation to each other. A construction is not so much built as composed. The making of a world seems such an indefinite act, subject to delay, so productive of disagreement, so overlapping with other offices, that it seems that the architectural office is an ideal space for worldbuilding as well as for architectural building. Indeed, perhaps that is what it does.

Central to the argument is that there is an infinite plurality of worlds. This claim introduces a dramatic effect regarding how we usually describe the world. What we have been calling 'connections' contributes to what? Certainly not the look of it unless it is showing you something; certainly not its argument unless it is debating with you. Put pompously, it is our culture. Worlds are where culture hangs out. Culture is the worlds we make.

This conception of worlds has implications for philosophical theories and the objects that sail under their flag. One issue is the object, what it is and what it doesn't get to be. In the concept of a world the object ceases to be a primordial thing that is the real basis of that world; it just becomes another connection. The overall construction doesn't wait there for a translation, or for an intellectual synopsis to reveal its meaning or its significance. A world seems more like a work by artists like Jasper Johns or Robert Rauschenberg. It will not accept a distinction between things and representations. The domain of representation has passed away into the administration of things.

Once the sky was dying it would only be a short while before God was declared to be dead

Historians of ideas will rush to qualify this by insisting that the world is a category behind which is a distinct concept. Middle Eastern texts mention 'heaven and earth'. Early Greek is the same; when Homer wishes to speak of all things, he gives a list. Hesiod refers to 'all things', to *panta*. But it was one of the possibilities of the Greek language to transform an adjective into a noun. The Greeks chose the word *cosmos* whose meaning included order as well as ornament (as we use the term cosmetic). The cosmos included everything including itself, but was separated from the human subject. They were reconnected by the fact that the human subject was to imitate the cosmos. In this way the cosmos became a norm of order; cosmology was a moral order which humans should imitate. That was how architecture could be seen as being governed by mimesis. The practice of imitation was a moral practice. The world was also 'worldly', and its study instructed us how to act. It was on this condition that morality and justice could be seen as being 'objective', that the subject related to the world by imitation.

This situation began to erode when the sciences began to grow out of the conceptual space of the 'world'. Astronomy began to leave behind the structure of the cosmos. At the end of the 18th century there was a poetic fear that the sky and the firmament were dying, displaced by the new indifference of astronomy and infinite space. Once the sky was dying it would only be a short while before God was declared to be dead. The world survived this disenchantment, but only by growing much smaller and becoming pluralist.

At one level, the world had become a category to help us understand an object. Recognising that an object may well be determined in part by its context, the world is a category which suspends the difference between the object and the context by affirming that together they constitute a world. Anything might be a world; anything we want to understand.

Ayham Jabr, *Fantasy of Life*, 2018

An important factor of imagining worlds is time. Not merely the theme of time, but the time spent thinking over a world, modelling it if only in the mind's eye. Duration allows details, layers, histories to evolve; requirements of the neurotic object. Geology shifts and weather systems develop. Biology warps around divergent paths of evolution. Backstories graft onto such worlds, myths of creation.

This is why the architectural model and its elaboration can benefit from the idea of 'world'. It removes the question of the model as having some definitive relation with what is being designed. It doesn't have to look or feel like a measured representation. It doesn't 'represent' at all. It models, in the sense that a potter models the clay. It places construction in the place of representation. Elements of a world are designed from other worlds. Intuitively, we can immediately see that the category of world is a much better point of view from which to study the connections between buildings. It neither makes style into a language of translation, as when we say we recognise the Baroque, nor does it suggest that the building is the autonomous creation of an architectural author.

By Design

Proposing that these questions be answered by attending to the concept of world, Nelson Goodman's book *Ways of Worldmaking* (1978) stipulates that the book itself must be taken as part of the production of its own world. In the book the world of philosophy is given a decisive shift. He outlines typical ways of making a

Ayham Jabr, *The Microswitch of Life*, 2017

left: Theodore Sturgeon's 1941 novelette *Microscopic God* describes a scientist able to create synthetic life in the laboratory. His miniature creatures, called Neoterics, evolve more swiftly than humans and achieve technological discoveries that the scientist can present as his own. The story ends with a power struggle after the scientist attempts to take over the world using a Neoteric power source. Such stories have precedents including Johann Wolfgang von Goethe's 1832 *Faust, Part Two* and the creation of homunculi.

The model has been displaced as an image or object which imitates the designed object-yet-to-be. In a sense everything becomes a potential model

Ayham Jabr,
Discovering Life,
2017

Opposite: Jabr here evokes the ocean explorer Jacques Cousteau. Maritime worlds permit the fantasy of worlds hidden within our own. Jules Verne's *Twenty Thousands Leagues Under the Sea* (1870) followed Captain Nemo's plan to establish a rival underwater civilisation. The nested world theme continued in Verne's sequel, *Mysterious Island* (1875); the *Nautilus* in both books being a self-contained seed world. The working out of so many details – ocean farming, submarine mechanics, distillation of drinking water – all contribute to Verne's worldmodelling project.

Ayham Jabr,
The Lord of Failure,
2019

Jabr brings his collage worlds to comment on the world, his world within Syria, in this work with its depiction of warfare and migrant camps. For Jabr, all the villages are really linked to his observations of Damascus, not unlike how Italo Calvino describes 55 supposedly fictional places in *Invisible Cities* (1972) when they all are really different perspectives of Venice. The fictive displacement allows for a wider analysis.

world, once established that its making is always a re-making. The first is Composition and Decomposition, how we put together and take apart stuff. Composition gets a very Constructionist account, for he equally insists that construction proceeds through a process of identification (not identification with something but identifying it as something), the organisation of stuff into entities and minds. Nothing is fundamental or prior to anything 'apart from a constructional system'. The model has been displaced as an image or object which imitates the designed object-yet-to-be. In a sense everything becomes a potential model. With Goodman's account it would be the 'name' model, meaning that it is nominated as the name of a special relation to the object.

Inevitably, he has to account for a philosophical interest in truth and in knowledge. It is clear that he thinks that the problem is not that they have no role, but that neither category is a central question and that the answers they offer are splendidly ordinary. The grandeur, the antiquity, the gravity of philosophical questions is shelved (indeed his world would need a lot of shelves) and the answers read somewhat like a smart construction manual. It has a democratic feel in that he is not providing a philosophical answer to a question but giving an account of what people do with objects, one that draws from the dispensations of philosophy. But he leaves no doubt that the democratic arena is also an invitation to leave no philosopher colleague unturned.

So the question of truth can arise, but only when a version of a world is verbal and made up of statements. Then the truth cannot arise as a test of statements in respect to the 'world'. Because there isn't one; there are only worlds. In Goodman's terms 'a version is taken to be true when it offends no unyielding beliefs, and none of its own precepts'.[2] Truth is less like an ultimate value and more like a 'fit'. The demotion is even clearer in the following statement: 'Truth, far from being a solemn and severe master, is a docile and obedient servant. The scientist who supposes that he is single-handedly dedicated to the search for truth deceives himself … He seeks system, simplicity, scope, and when he is satisfied on these scores he tailors truth to fit. He as much decrees as discovers the laws he sets forth, as much designs as discerns the patterns he delineates.'[3] This adds up to redesigning the space of architectural work. It consists, at a general level, of a world in which the design functions. A 'world' dispenses with the awkward distinction between an object and its context, and models can emerge which are not tied up in an anachronistic attempt to make an object like the final object, only smaller. ⌓

Notes
1. Nelson Goodman, *Ways of Worldmaking*, Hackett Publishing Company (Indianapolis, IN), 1978.
2. *Ibid*, p 17.
3. *Ibid*, p 18.

Text © 2021 John Wiley & Sons Ltd. Images © Ayham Jabr

Pascal Bronner and Thomas Hillier

FleaFollyArchitects,
Modern Prometheus,
Jerwood Space, London,
2014

This 'grotesque shrine to the digital age' holds inconceivable amounts of information, transforming it into an infinite electronic archive. It is sited within the barren lands of the Arctic Circle, the location of the last sighting of Mary Shelley's creature, where the desolate, glacial climate is used to cool this infernal creation.

Handmade Worlds

Constructing an Inhabitable Modelscape

Co-founders of FleaFollyArchitects, **Pascal Bronner and Thomas Hillier** specialise in the disarming effect of distorting the standard rules of architectural scale. Their propositions consist of wondrous, complex models often based on fictitious stories that they take into the outside world to contribute to the story of the city.

As a practice, FleaFollyArchitects was born through modelmaking; even the name is a play on notions of inhabitation, the ambiguities of scale and a sense of mystery. Over time the studio has used models to formulate a methodology that explores how the agency of modelmaking can be used to bring architectural microcosms to a wider audience. Can the boundaries of architectural modelmaking be pushed beyond their traditional limits into a realm of miniature architecture? And if so, could this architecture be inhabited through the mind, and even the body, creating a building and not a model, or maybe a hybrid of both?

Furthermore, could an architecture be built entirely by hand within the confines of the architects' workplace? Could we go beyond being the drafters of buildings and become the builders of buildings or even worlds? And how can these ideas be propelled into architectural practice, with real clients, sites and constraints? At this point are they still classified as models?

Narrating a Model World

FleaFolly has always employed narrative to explore, discover and invent new worlds, particularly through modelmaking, turning stories into three-dimensional, spatial constructs. In 2012, alongside seven intrepid explorers (a group of our first-year students from London Metropolitan University) we developed our first large-scale model, *Grimm City*. To celebrate and mark the 200th anniversary of the first publication of *Grimms' Fairy Tales*, on the outskirts of the Black Forest in Germany we planned and built a new metropolis, a meticulously handcrafted miniature Cityscape based on the imaginations of the Brothers Grimm.

Intentionally disregarding scale, *Grimm City* was designed not as a typical architectural model, but as a miniature architecture, or modelscape, that represents an entire world. While most architectural models scale a building at one concurrent scale, the abundance of different scales in *Grimm City*, alongside familiar objects such as bowler hats and test tubes, root it firmly within the impossible. The model aimed to embody the transitory and intangible nature of architecture that in theory will not only mimic, but also replicate spatial moments so that the viewer can wholly consume its infinite landscape. The observer becomes a traveller, a denizen of the city, more than an onlooker, but an inhabitant through the imagination and to some degree the body. Could this concept be pushed further? In *Modelling Messages: The Architect and the Model* (2005), Karen Moon suggests that 'The reduced scale of most models gives rise to a perennial problem: they are too small to enter. Too often the viewer's experience is restricted to the building's external form. Only with the imagination (or a little peeping through the window) can we access the interior.'[1] What followed, in quick succession, were two further models that built upon the notions explored in *Grimm City* that aimed to disrupt the 'restrictions' between the inside and out, blurring the threshold of the model and of the inhabitable.

Inhabiting Model Worlds

The first was a commission by the Jerwood Gallery in London, entitled *Modern Prometheus* (2014). This 4.5-metre (15-foot) tall monument, built entirely within the studio, explored how the information age has turned our lives into a digital frenzy of passwords, codes and social networks. Using themes from Mary Shelley's *Frankenstein*, this 'Technological Tower of Babel' explores, interprets and manifests the challenges that we as humans face in the information age.

FleaFollyArchitects,
Grimm City,
Design Museum,
London,
2013

This architectural satire foretells a future state run by creatures with Grimm-esque attributes of gluttony and greed taken perversely out of context. Whilst lampoonist and satirical, this architectural fairytale draws uncanny parallels with what many believe is the way cities are run today.

FleaFollyArchitects,
Modern Prometheus,
Jerwood Space,
London,
2014

Rings of accommodation made from black laser-cut card wrap and blanket the tower, creating an inhabitable layer of insulation that keeps the internal core at an optimal temperature.

Building on Moon's notion of 'peeping', glimpsing the inside of an architectural model through a window or opening often does not reveal much more than one could have imagined looking at its external form. The *Modern Prometheus* was designed to be as permeable and thus as accessible as possible, and although not physically inhabitable, the sheer height, scale and levels of intricacy created an imposing form. Conversely, the formal nature of the 'tower', no matter how intricate and alien, gave the viewer an immediate sense of an architectural typology, while the intentional use of duplication and a less ambiguous scale, in this instance, created a form of barrier between human and model. This meant the piece did not simulate human occupation as well as imagined.

Using what was learnt from the previous two models, the final model in this series, again built with a group of students in the Black Forest, aimed to further question and explore how inhabitable a model really could be. In 1974, Nobel Prize-winning writer Elias Canetti wrote *Der Ohrenzeuge: Fünfzig Charaktere*, which translates into 'Earwitness: Fifty Characters'.[2] His novella portrays 50 contemporary life exemplars, each a paradigm of a certain type of behaviour. Although written over 40 years ago, it is still seen as incredibly relevant and comparable to many of today's societal traits. From 'The Blind Man' whose entire life exists through photographs or 'The Name-Licker' who, above all else, craves the attention of celebrities. Inspired by this, in 2015 FleaFolly created its own *Earwitness*, a fictitious village where each of the 50 characters dwells within their own home and garden plot, each of which tectonically represent their traits as viewed in today's society. Each plot has been custom-designed, meticulously handcrafted in white card, and is situated within an arched portal that forms the outer perimeter of the village.

Inspired by 'Newton's Cenotaph', Étienne-Louis Boullée's visionary unbuilt work of 1784, the entire village is housed in a giant temple-like structure over 3

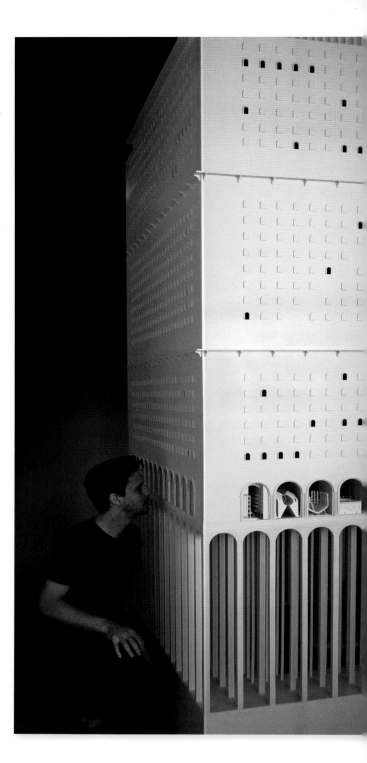

FleaFollyArchitects,
Earwitness,
Black Forest,
Germany,
2015

above: The pure-white monolith gives the viewer a variety of ambiguous and strange glimpses into its more delicate interior, attempting to capture and captivate the imagination of the observer.

opposite: Once inside this hidden universe constructed of paper and card, the viewer, now resident, is surrounded by tectonic renditions of Elias Canetti's 50 characters, with the viewer's own head forming the village square at its centre.

Each plot has been custom-designed, meticulously handcrafted in white card, and is situated within an arched portal that forms the outer perimeter of the village

FleaFollyArchitects,
The Listening Tower,
Studley Royal Water Garden,
Ripon, North Yorkshire,
England,
2018

opposite: Taking inspiration from both the long-lost 18th-century Bathing House that once stood in the garden, and a 17th-century Japanese water harp, FleaFolly built a temple that contains a giant terracotta vase that acts as an echo chamber to create an ever-lasting, finely tuned melodic splashing sound that is amplified through a series of copper 'listening ears'.

FleaFollyArchitects,
Hak Folly,
St John's Gate,
Clerkenwell,
London,
2016

below: A 4.5-metre (15-foot) high temple of timber located within the historic St John's Gate, *Hak Folly* was a stacked structure using off-the-shelf floorboards that was held together through gravity and compression alone with no fixings used within.

metres (10 feet) in height. It can be viewed through the aforementioned arches alongside a series of windows that punctuate the façades. These moments give the viewer very intentional, and to some degree detached views of the internal workings of the village, while ensuring that the scale (which is never specified) remains uncertain. This ambiguity is further enhanced by the entire piece being white throughout, ensuring one's own views and memories on materiality are inconsequential.

Yet it is not until the forest of long, thin columns that hold the structure in place open, revealing a small, mobile seat, that the true nature and physical relationship to the model becomes apparent. The seat slowly moves the viewer into the model's underbelly where they become intrinsically linked to its internal universe by placing their head up through into the centre of the village. In a sense, one inserts one's head both physically and psychologically into the model. The viewer crosses a threshold between the spectator's space and the *Earwitness*'s space with the aim of transforming the act of viewing a model to that of actually travelling to this miniature architectural world.

Modelling In Practice

Can these constructs made of paper, card and wood truly be seen and experienced as architecture? We do not know, and are not sure that we need to know, and this is part of their inherent mystery. FleaFolly most certainly sees the work as a form of worldbuilding on a miniature scale, with these models and the ideas behind them being incredibly influential on our evolving practice and larger works. This is evident in two further commissions, *Hak Folly* (London, 2016) and *The Listening Tower* (Ripon, North Yorkshire, 2018). Both were made entirely by us in our studio, and we see them firmly as large-scale models rather than pieces of architecture. Their use of materiality and internal complexities that are not immediately apparent, alongside an intentional vagueness of how they are actually to be used, continue to blur the boundaries between what is a model and what is architecture. A boundary we continue to blur. ⌀

Notes
1. Karen Moon, *Modelling Messages: The Architect and the Model*, Monacelli Press (New York), 2005, p 62.
2. Elias Canetti, *Der Ohrenzeuge: Fünfzig Charaktere*, Hanser Publications (Munich), 1974.

Text © 2021 John Wiley & Sons Ltd. Images © FleaFollyArchitects

Chad Randl

Remo
Home as

delling Cosmos

Detail view of fireplace mantel,
Young House,
Atlanta,
Georgia,
1997

Fireplaces, and the floors and walls surrounding them, are a primary site for the accumulation and arrangement of expressive objects. This example, recorded by the United States National Park Service's Historic American Buildings Survey, has likely been reconfigured over time.

House refurbishment, renovation or simply rearranging the pictures on your walls or ornaments on your shelves is worldbuilding. **Chad Randl**, Art DeMuro Assistant Professor in the Historic Preservation programme at the University of Oregon, takes us on a psychological trek via our domestic environments to reveal that we worldmodel consistently and perennially at home.

Wife of local merchant arranging bric-a-brac in her home, Concho, Arizona, 1940

A United States Farm Securities Administration photo by Russell Lee. In times of scarcity, such as during the Great Depression, a room can be remade by reorganising displayed objects of meaning.

In 1958, French philosopher Gaston Bachelard wrote that 'our house is our corner of the world ... our first universe, a real cosmos in every sense of the word'.[1] It is in and around the home that we ward off that sense of impotence that arises from the infinite, where we feel some power to effect change. Every alteration of the walls and floors and the spaces and things within is an effort to remodel the world into something uniquely ours.

Rearranging as Remaking

At its most modest, remodelling can include reordering the objects that surround us – our furniture, wall hangings and things that decorate shelves and sills. The collection and assembly of objects has long been a way to personalise our dwellings and identify who we are to ourselves and to others. We establish associations through statuettes and objets d'art (Hummel figures and japonisme), and we recall our travels with gift-shop plates, snow globes and spoons. We commemorate lost loved ones in portraits and photos.

As occupants add years and as their tastes and means evolve, these settings evolve accordingly – some things are added, some disappear. Anthropologist Victor Buchli has argued for the value of studying such assemblies and juxtapositions that reveal conditions of 'radical discontinuity', 'undecidability' and 'conflict'.[2] Posters of favourite bands come and go from the walls of the adolescent bedroom. Furniture is reconfigured to escape ennui, accommodate new living patterns and occupants, or to clear paths for ageing bodies. Such peripatetic habits are aided by wheeled and ball-bearing casters that became increasingly common in the latter 19th century. Rearrangement is still considered a best first step when dwellers are 'feeling stuck' or, in an age of pandemic and quarantine, they have 'too much time on [their] hands'.[3]

Purging is also remodelling. In 1897, at the end of an era that celebrated ornamental excess, novelist Edith Wharton wrote: 'Decorators know how much the simplicity and dignity of a good room are diminished by crowding it with useless trifles. Their absence improves even bad rooms or makes them at least less multitudinously bad. It is surprising to note how the removal of an accumulation of knick-knacks will free the architectural lines and restore the furniture to its rightful relation with the walls.'[4] Here, objects that once held meaning turn to detritus and bric-a-brac. Groupings carefully arranged become clutter, forsaken in the name of simplicity and the clean line. Author and 'tidying expert' Marie Kondo has recently promised life-changing benefits for asking whether or not each possession sparks joy and disposing of those that do not.[5]

Rearranging and purging may be as far as the tenant can go in shaping their cosmos. Owners can do what they want within the bounds of building codes and budget. The deep-pocketed remodeller or the owner of the house deteriorated almost to ruin may strip it back to studs. Scenes of sledgehammer-wielding hosts peeling a house back to its 'good bones' are fixtures of home-improvement television programmes. Substantial remodelling campaigns often introduce new services and amenities such as running water, electricity and heating systems. Over the past century, projects have focused on opening-up older floorplans featuring partitioned kitchen, dining and living rooms. During the same period, drawing from efficiency and motion studies first applied in factories, domestic engineers called upon homeowners to relocate sinks, cupboards and cookers to save steps and reduce unnecessary movement.

Charles and Ray Eames,
Eames House,
Pacific Palisades,
California,
1949

Ray and Charles Eames were avid collectors of vernacular and folk-art objects that they would display in the living room of their self-designed home. The couple regularly used the home as a backdrop for promotional photoshoots featuring their own furnishings and other projects. The house illustrated the tension between a sparse, machine-made modern aesthetic and one that appeared lived-in, personalized and homely.

Layers Upon Layers

If gutting is like starting over with a more or less blank slate (the tabula rasa), a less intrusive tactic – layering – evokes the palimpsest, the reused parchment that bears traces of earlier writing. Construction itself is a process of layering. Builders lay finish floors on subflooring that sits atop floor joists; plasterers apply base, skim and finish coats to lath that itself bridged wall studs. Remodelling extends these practices into periods of occupancy, and layers allow those practices to be read over time. Layering of wall, floor and ceiling finishes is prompted by two primary forces: changes in fashion and to escape burdensome (or too-long deferred) maintenance.

On the interior, coats of paint, varnish and washes, dozens of layers thick, embed a stratigraphy of once-stylish tones that give way one to the next. Strata of wallpaper may be interspersed, similarly revealing formerly popular decorative trends. Beat-up hardwood floors are refreshed with paint or stain then covered with linoleum, which in turn is hidden beneath wall-to-wall carpeting. Stair rails are boxed in with two-by-fours and drywall to streamline the interior and conceal loose balusters or newel posts. Cracking plaster is covered by painted canvas, cotton sheeting or wallpaper, and when it sags and buckles is cocooned behind wall board.

Upstairs apartment in a commercial building, Girard Avenue, Philadelphia, Pennsylvania, 2010

Remodelling campaigns may reveal a history of wall finishes and previous interventions dating back a century or more. Here, at least five layers of wallpaper and multiple layers of paint (some over, some under the wallpaper) provide a chronology of styles, taste preferences and cultural moments. Hopalong Cassidy was a fictional cowboy featured in books, films and television programmes.

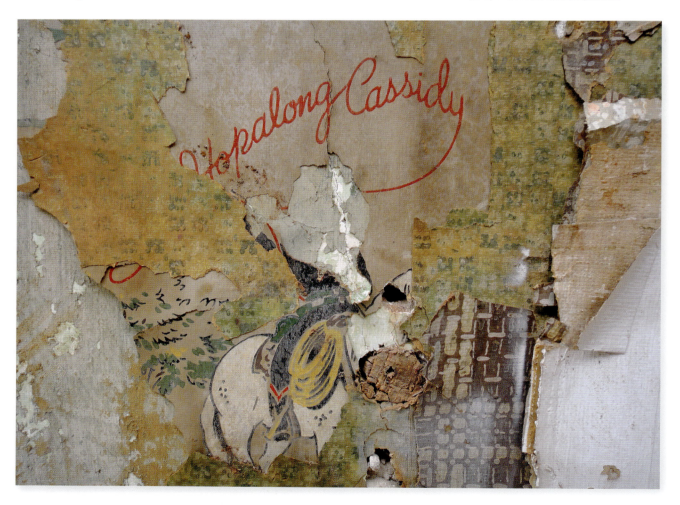

1850s early Willamette
Valley house,
Oregon,
2003

When white settlers established homesteads in North America, they often built log-hewn structures. As the family became more established they upgraded the house by sheathing it in wood clapboard, a more weather-resistant and dignified finish. In time, successive occupants, tired of repainting and repairing, might cover this siding with asphalt shingles, cement asbestos siding, or aluminium or vinyl siding.

Exterior surfaces are also layered with paint and other materials. The owners of early log houses sheathed them in wood siding as soon as they could afford it. Building product manufacturers developed a range of new wall treatments that could be easily applied atop older layers, by professionals and increasingly by amateur do-it-yourselfers. Rolled asphalt sheets with a pressed-brick appearance could be tacked directly onto older surfaces. Formstone and shotcrete were concrete-based facing materials regularly applied over existing façades, the former poured into moulds to resemble stone, the latter applied with a pneumatic gun in a smooth or pebbledash finish. Installing new roofing on old was said to save labour and provide additional weather resistance. A 1926 article in *Popular Science Monthly* advised readers to 'Leave old shingles on the roof, unless too curled. Cover with new material. Let outside wall covering stay beneath new siding, shingles or stucco.'[6]

Dennis Maher,
City Chancel,
2017

Artist Dennis Maher's work examines inhabited spaces subject to regular reconsideration, remaking and redefinition. His wall hanging ruminates on buildings as layered artefacts. Inside and outside are juxtaposed and blurred. Exterior materials cover interior walls; interior surfaces seem to sheath exterior walls.

Seeking Meaning in Remodelling

What can layering and rearranging say about the domestic worldbuilding that is inhabitation and remodelling? Is the piling of layer upon layer a means of constructing a hardier, more resilient shell to keep the world beyond at bay? Do all these membranes provide insulation from the intrusion of wind, wear, obsolescence and other realities we do not want to face? Do they weigh us down with the mess and the mysteries enfolded within, of past owners and occupants, of who knows what went on here? Are the objects we organise and reorganise the physical manifestation of the thoughts in our minds that we constantly reconfigure to try to make sense and find comfort? Does the rearranging of furniture, instead of autonomy and initiative, signal futility and insubstantial change, evoking the now hackneyed phrase of shuffling the deck chairs on the Titanic?

In *The Poetics of Space*, Bachelard undertook what he called a topo-analysis of the home to identify a 'topography of our intimate being'.[7] Such topos have an outward as well as an inward character. Inner thoughts and life ways, expressed through the shape of our things, are projected wishfully into a world over which we have little control. Bachelard drew upon psychiatrist Carl Jung's comparison of the human psyche to a stacked structure built up over time with a once-occupied cave at the base, over which were accreted an ancient Roman foundation, a 16th-century ground floor built from an 11th-century tower, topped by a 19th-century upper floor.[8] But remodelling practices demonstrate a different form of accretion. The remodelled home is less the 'vertical being' that Bachelard imagined rising upward, and more one that grows inward and outward, reinforcing our sense of self as it incorporates other pasts.[9] It is in these tentative, incremental, ephemeral changes, the transience of rearrangement, the building up of thin layers one atop the next, that the dweller reshapes their universe. Remodelling is worldmodelling. ⌂

Notes

1. Gaston Bachelard, *The Poetics of Space* [1958], trans Maria Jolas, Beacon Press (Boston, MA), 1994, p 4.
2. Victor Buchli, *An Archaeology of Socialism*, Berg Publishers (New York), 2000, p 5.
3. Beth Bruno, 'If You're Feeling Stuck, Rearrange the Furniture', *Medium*, 29 January 2020: https://medium.com/@bethbruno2015/if-youre-feeling-stuck-rearrange-the-furniture-f34a730aa04; and Ronda Kaysen and Michelle Higgins, 'Rearranging the Furniture to Pass the Time', *New York Times*, 12 April 2020, p 5.
4. Edith Wharton and Ogden Codman Jr, *The Decoration of Houses*, Charles Scribner's Sons (New York), 1897, p 185.
5. Marie Kondo, *The Life-Changing Magic of Tidying Up: The Japanese Art of Decluttering and Organizing*, Ten Speed Press (Berkeley, CA), 2014. The quote is from Kondo's website: https://shop.konmari.com/pages/about.
6. John R McMahon, 'A Dozen Tips on Remodeling Old Houses', *Popular Science Monthly*, 108 (5), 1926, p 28.
7. Bachelard, *op cit*, p xxxvi.
8. CG Jung, *Contributions to Analytical Psychology*, trans HG and Cary F Baynes, Harcourt, Brace (New York), 1928, pp 118–19.
9. Bachelard, *op cit*, p 17.

Text © 2021 John Wiley & Sons Ltd. Images: pp 56–7 Library of Congress, Prints & Photographs Division, HABS Reproduction number HABS GA,61-ATLA,66–13; p 58 Library of Congress, Prints & Photographs Division, HABS Reproduction number HABS GA,61-ATLA, 66–13; p 59 Photographer Leslie Schwartz © Eames Office LLC (eamesoffice.com). All rights reserved; p 60 ©Tracy Levesque, 2010; p 61 photo Pacific Northwest Preservation Field School; pp 62–3 © Dennis Maher

Theodore Spyropoulos

Everything You See is Yours

Minimaforms (Theodore and Stephen Spyropoulos),
Emotive House,
SCI-Arc Gallery,
Los Angeles,
2016–

Emotive House is a framework that collects social data from social-media platforms, and urban data from API streams, to enable behaviours in this proposed mobile-living unit. The house processes the data and assigns them into cells or units of interaction, reorganising its internal structure and collective self-organising as a new model of urbanism.

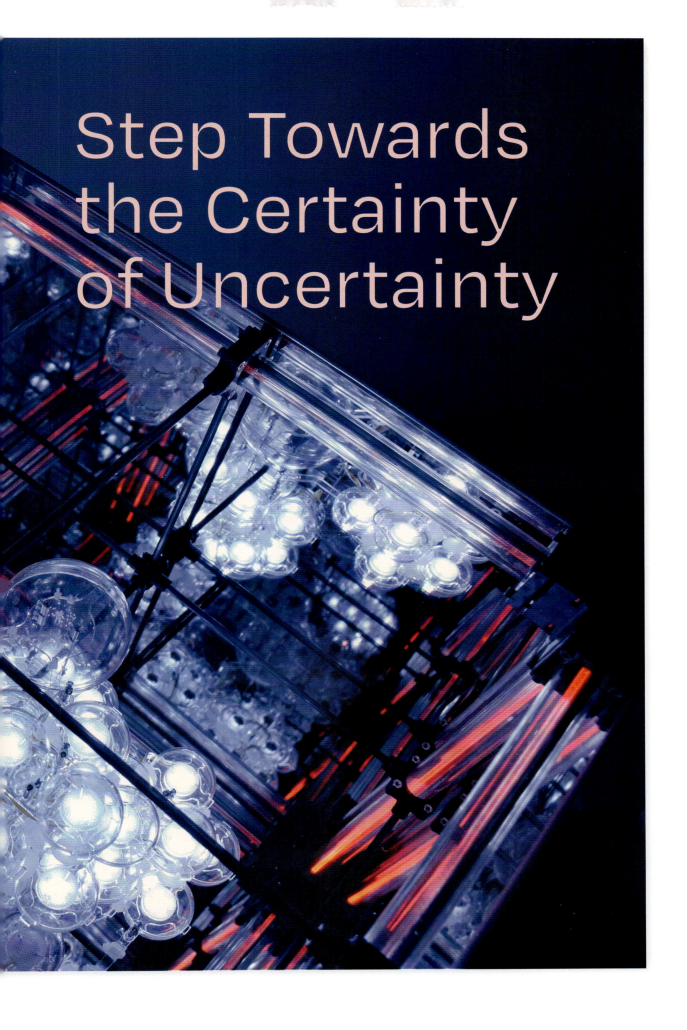

Step Towards the Certainty of Uncertainty

Director of the Architectural Association Design Research Lab (AADRL) and co-founder of Minimaforms in London, **Theodore Spyropoulos** describes the necessity to construct artificially intelligent environments that explore our technological sphere so that we may better understand and actively participate in the emerging complexities.

In 1969, George Spencer-Brown published his seminal book *Laws of Form*,[1] an attempt to straddle the boundaries between mathematics and philosophy in which he declared: 'Draw a distinction and a universe comes into being.'[2] In this one statement the paradoxes of worldbuilding and our relationship to it are outlined. If our understanding of the world is ours, then it remains without action inaccessible to others. The 'world', rather than something shared and understood, is plural, situated and in continuous formation. Worlds within worlds understood through a cosmology of observations. This conceptually outlines the philosophy of radical constructivism, expanding on a second-order cybernetic understanding that places the observer in the world of its observation. The observer here is understood as human or nonhuman/machine. Heinz von Foerster, the famed cybernetician and director of the Biological Computer Laboratory at the University of Illinois Urbana-Champaign, expressed this when he stated: 'Objectivity is the subject's delusion that observing can be done without him.'[3] The certainty of all forms and their representations therefore are cast into doubt. The only certainty remains uncertainty.

Architectural discourses today shy away from uncertainty and complexity. The current lines of thought champion representational models of historical appropriation, dogmatic styles or an all-out rejection of architecture having meaningful social or political agency through design. These approaches reinforce the normative with new narratives and representations without critically addressing how architecture can actively participate in the world. In contrast to this, we live in a technologically interconnected social and ecological sphere that has made us hyperaware of the magnitude and complexities of the challenges today. By necessity, if contemporary design is to remain relevant it must shift from the finite representational models of practice towards real-time collaborative ones. The shift conceptually is to move from 'models of' towards 'models for'. Enter design.

Modelling Paradox
The role of the model today must be operative and prototypical in nature. Rather than illustrating ideas, models should offer us the possibility to engage and respond to the information-rich matrices that influence our everyday. Gordon Pask spoke of this as the paradoxical nature of what appears tangible and intangible within architecture. In his paper for the *Proceedings of the Seventh European Meeting in Cybernetics and Systems Research* entitled 'The Architecture of Knowledge and Knowledge of Architecture', he stated that 'Increasingly we are immersed in this environment and dominated by its technologies; soon, before a point of sheer engulfment, humankind will perceive this environment, which increasingly determines the orientation and the ethos of humankind. We live in a world shaped by, and in the image of.'[4] Pask wrote these words in 1984, and though not Orwellian in spirit spoke to the necessity

Minimaforms (Theodore and Stephen Spyropoulos),
Emotive City,
London,
2015

The *Emotive City* uses everyday interaction as a fundamental communication model to create ecologies of mobile and self-structuring habitual environments. This augmented environment constructs a new nature of human–machine interactions structured through behaviours that challenge the fixed and finite masterplan as an urban mode.

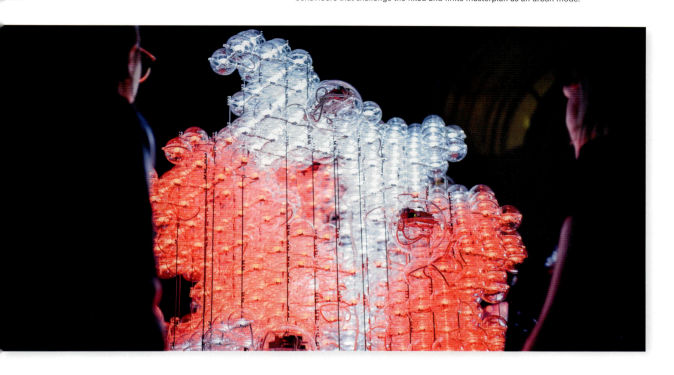

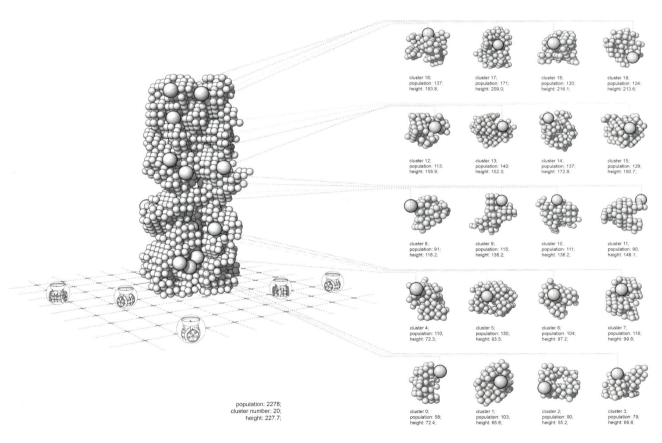

Minimaforms (Theodore and Stephen Spyropoulos),
Emotive City,
London,
2015

Generative body plan organisations are computed through rule-based interactions, synthesising information streams. Formations are constructed in real time, moving from dynamic to stable states as the model iterates emotive stimuli, localised through cluster interactions.

Architecture must re-conceptualise the role that space as medium can play today by rejecting the historical crutches that reinforce modern and postmodern rehashing

Minimaforms,
Petting Zoo,
'Digital Revolution',
Barbican Centre,
London,
2014

Petting Zoo is a speculative robotic environment populated by artificial intelligent creatures that have been designed with the capacity to learn and explore behaviours through interaction with participants. Within this immersive installation, interaction with the pets fosters human curiosity and play, forging intimate exchanges that are emotive, evolving over time and enable communication between people and their environment.

Minimaforms (Theodore and Stephen Spyropoulos), *Memory Cloud*, Trafalgar Square, London, 2010

Memory Cloud is a participatory framework based on the ancient practice of smoke signals – one of the oldest forms of visual communication. Fusing ancient and contemporary mediums, it creates a dynamic hybrid space that communicates personal statements as part of an evolving text, animating the built environment through conversation.

in both theory and practice to actively examine these technologies that shape our lives.

Fast-forward to today, and it can be argued that not engaging actively with these technologies and their byproducts renders architecture both intellectually and operationally in a position of obsolesce. The knowledge of architecture necessitates swimming in the challenges posed by this new world through radical experimentation and response. Architecture must re-conceptualise the role that space as medium can play today by rejecting the historical crutches that reinforce modern and postmodern rehashing. Technologies released into the world are in the world, and for society to not fall subservient to them demands that we deeply understand them. This understanding is the only means to offer other possibilities for their implementation than those that may have realised them.

Models should be understood as open frameworks and platforms enabling shared and collective participation. 'Design then must be considered as durational, real-time and anticipatory exploring human with human, human with machine, and machine with machine communication. The challenge posed is how designers can construct environments that enable curiosity, evolve and allow for complex interactions to arise through human and nonhuman agency. Attention here is placed on behavioural features that afford conversational-rich exchanges between participants and systems, participants with other participants and/or systems with other systems. This evolving framework demands that design systems have the capacity to participate and enable new forms of communication. Beyond convention architecture moves towards features that are life-like, machine learned, and emotively communicated.'[5]

The architectural model conceived in this manner affords us the opportunity to see architecture as an active environment for spatial interfacing and stimulus for participation. Through bypassing generalised aspects of systemic practice we can construct novel models of interaction that evolve through our collective interfacing. Architecture should foster intuitive and behavioural attributes of communication, giving agency to the collective through their action as influence.

Evangelos Polykandriotis,
Giulia Arienzo Malori,
Shiri Dobrinsky and Tao Yu,
Tropos,
Spyropoulos Design Studio,
Architectural Association
Design Research Lab (AADRL),
London,
2020

opposite: Tropos is a prototypical aerial infrastructural system that augments existing urban illumination and mobility models. Designed as a swarm-based population of intelligent agents, the project explores adaptive models of infrastructure and their implications in the urban environment.

below: Worldbuilding demands an active engagement that challenges normative methods of exploring architecture. The Design Research Lab seeks to augment our environment through the implementation of adaptive systems that are embodied and situated as infrastructures that evolve with us.

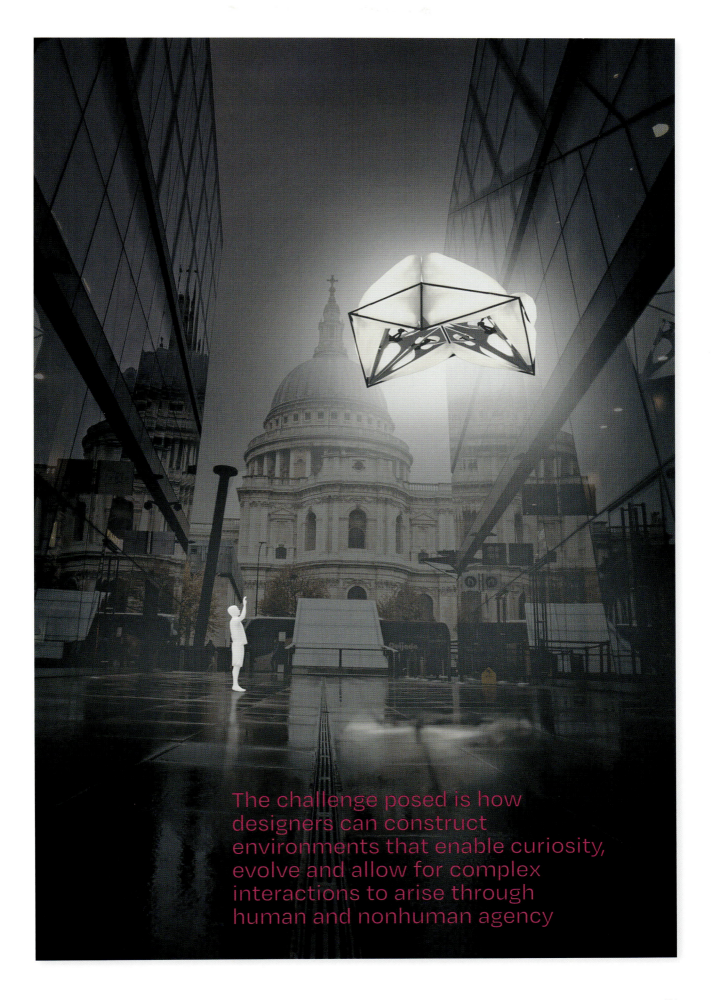

As technologies accelerate the collapse of space and time, architecture remains itself a form of resistance

Minimaforms (Theodore and Stephen Spyropoulos), *Of and In The World*, Somerset House, London, 2017–

Fifteen hundred glass orbs construct an inhabitable crystalline lattice that examines worldmodelling by exploiting physics, optics and our dynamic readings of space through our experience. The orbs use cosmological and celestial organisations of three offset spherical layers that interconnect each of them in a dynamic equilibrium. In conception and realisation, the work is a construct of the mind, demonstrating that what one sees is truly one's own.

The concept of worldmodelling moves beyond contemporary tendencies of observer systems that we find within artificial intelligence practices today that privilege reinforcement learning based on the habitual and known. In acknowledging the agency of human and nonhuman things that are situated within an evolving environment, we offer the capacity to see with the world of observation through possibilities of meaningful interaction. Architecture therefore needs to be an active participant moving beyond finite positions and passive observations towards a truly second-order cybernetic architecture that is both subject and interacting object within the world. This proposed AI would be understood as 'architectural intelligence', a learning environment challenging the simplistic binary conventions of the human and nonhuman or physical and digital through evolving spatial and interacting ecologies.

The concept of intelligence thus defined is not something attributed as a property to things themselves, but rather attributes arising between things, a product of the interface of their interaction. Beyond simplistic relationships that reinforce passive observation, this approach enables each participant to be a performer actively influencing and being influenced within its environment. It speaks to an architecture that learns and evolves with uncertain and at times latent understanding of the world in which it operates. Famed physicist Carlo Rovelli speaks to the necessity of uncertainty within the sciences when he says: 'The very foundation of science is to keep the door open to doubt.'[6] Beyond blueprint or masterplans, architecture today must argue for space as a medium of communication. The work of Minimaforms and the Spyropoulos Design Studio within the Architectural Association Design Research Lab (AADRL) in London demonstrates an approach to what constructing 'models for' may offer. Through enabling participation and affording agency we may find architecture must break its reliance on representation and notation and move towards the behavioural, offering an understanding of the deeper relationships between things. As technologies accelerate the collapse of space and time, architecture remains itself a form of resistance. ∆

Notes
1. George Spencer-Brown, *Laws of Form*, Cognizer Co (Portland, OR), 1994.
2. Bernhard Pörksen, *The Certainty of Uncertainty: Conversations on Constructivism*, Carl Auer Verlag (Heidelberg), 2002, p 34.
3. Quoted in Ernst von Glasersfeld, 'Declaration of the American Society for Cybernetics', in Constantin Virgil Negoita (ed), *Cybernetics and Applied Systems*, Marcel Decker (New York), 1992, p 3.
4. Gordon Pask, 'The Architecture of Knowledge and Knowledge of Architecture', in R Trappl (ed), *Cybernetics and System Research 2: Proceedings of the Seventh European Meeting in Cybernetics and Systems Research, Vienna, 24–27 April 1984*, North-Holland (Amsterdam), 1984, p 27.
5. Theodore Spyropoulos, 'Constructing Participatory Environments: a Behavioural Model for Design', PhD dissertation, University College London (UCL), 2017, p 7: https://discovery.ucl.ac.uk/id/eprint/1574512/.
6. Carlo Rovelli, 'The Uselessness of Certainty', 2011: www.edge.org/response-detail/10314.

Text © 2021 John Wiley & Sons Ltd. Images: pp 64–5, 67(t), 68–9 photos Theodore Spyropoulos; pp 67(b), 72–3 © Minimaforms; pp 70–1 Courtesy AADRL

Model & Fragment

On the Performance of Incomplete Architectures

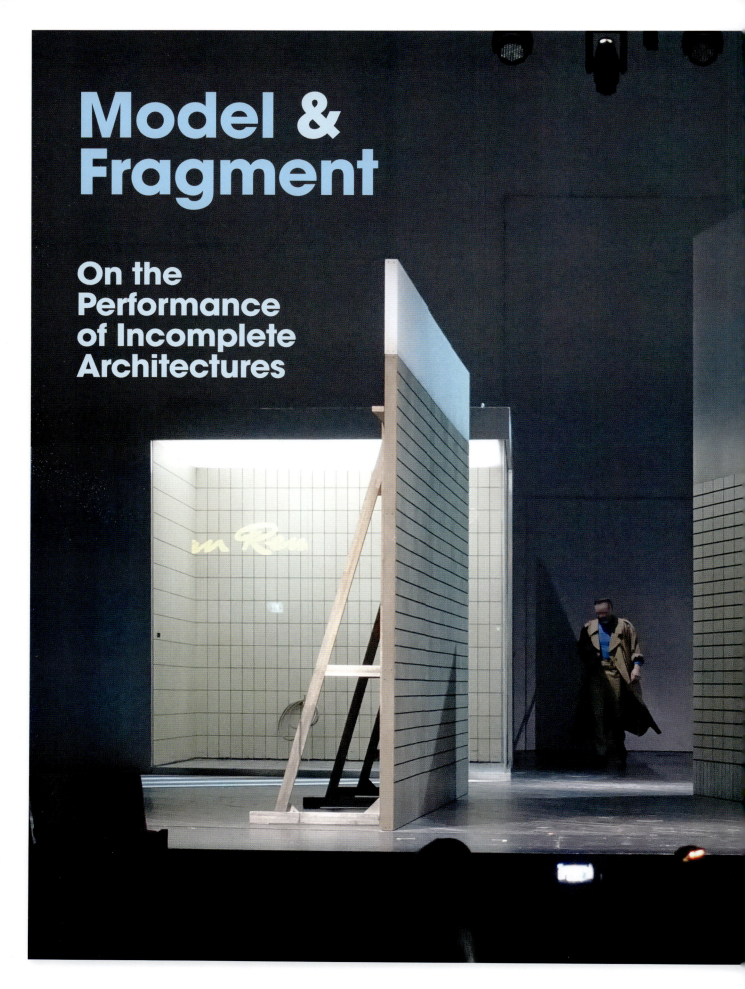

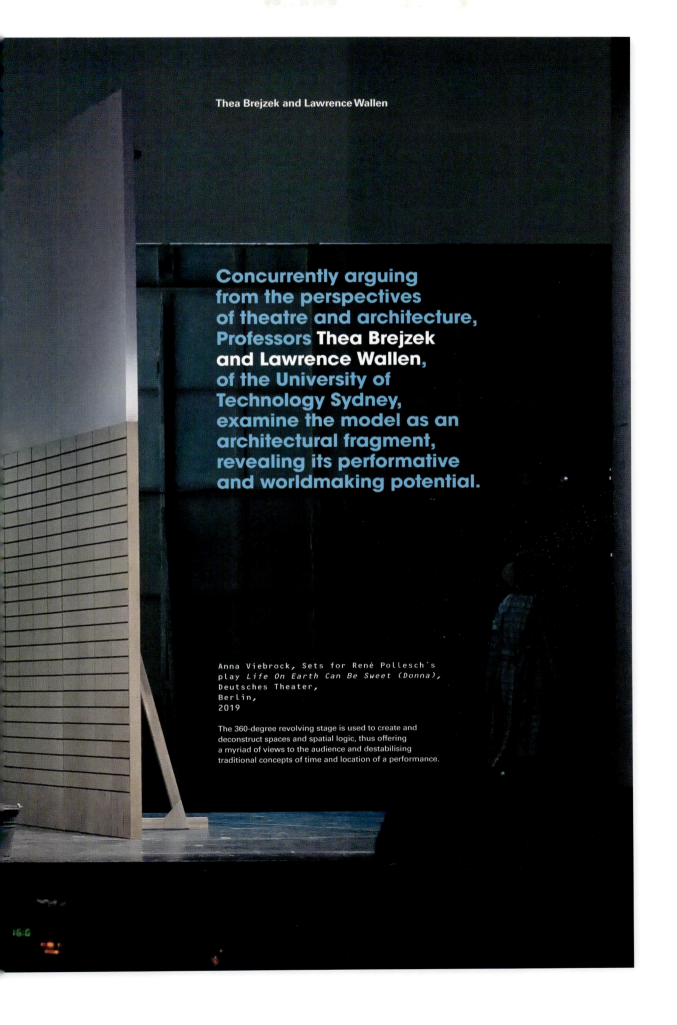

Thea Brejzek and Lawrence Wallen

Concurrently arguing from the perspectives of theatre and architecture, Professors Thea Brejzek and Lawrence Wallen, of the University of Technology Sydney, examine the model as an architectural fragment, revealing its performative and worldmaking potential.

Anna Viebrock, Sets for René Pollesch's play *Life On Earth Can Be Sweet (Donna)*, Deutsches Theater, Berlin, 2019

The 360-degree revolving stage is used to create and deconstruct spaces and spatial logic, thus offering a myriad of views to the audience and destabilising traditional concepts of time and location of a performance.

Architectural models are physical and conceptual instruments of the *cosmopoietic* (worldmaking) act that can comprise entire worlds. This performative and reality-constructing dimension of the model is intrinsically linked to processes of modelling, iteration and ideation. Theatre, as a physical practice and ephemeral art form, can equally be considered as a world maker in the creation and representation of constructed model worlds. Guiding these claims is the understanding that any model (mathematical, philosophical, mechanical, artistic) is always *a model of and a model for* – an idea, a concept, or a past, future or altogether experimental or utopic structure. What is a given is that the model's object status presents as the physical outcome of an idea.

Moreover, it is the desire for worldbuilding that is constitutive to the model's existence. In its duality of object and idea, the model retains a promise, imagined yet rarely fulfilled. The act of cosmopoiesis that describes the model's performative capacity to imagine and construct a world begins as 'making' in the studio and continues as 'representation' outside of its confines.[1]

The Fragment and the Whole
This article investigates the architectural model's creation, and reception aesthetics in an exceptional instance, namely when the process of worldmaking is deliberately disrupted and, in a twofold intentionality, constitutes an architectural fragment. This research understands the production of the fragment as an active process, taking its cue from Romantic literary critic and philosopher Friedrich Schlegel's 1789 provocation that the fragment activates and 'ferments' the human intellect and from its etymological origin, namely 'frangere' (Latin – to break, to shatter):

> A. Fragments, you say, are the actual form of universal philosophy. … But what can such fragments do for the most important and most serious matter for humanity, for the perfection of science? – B. Nothing but a spice of Lessing against intellectual rot … or even a fermenta cognitionis … .[2]

To Schlegel as to the Romantic movement overall, the (literary) fragment occupies a special place as both an isolated and complete small work of art. The fragment's unity, however, is not one of a harmonious consensus but rather presents a unique perspective of a 'chaotic universality'.[3] It is this intended dissonance within a unified single work that renders the fragment an evocative form beyond its origin in literature. The fragment is but a remnant of the shattered whole. The act of shattering is more than an aesthetic strategy; it is also a provocative act that produces specific meaning and constitutes, in itself, a performative and reality-constructing process. The instance of the model conceptualised, constructed and displayed as architectural fragment occurs in exhibition and on the stage today where the architectural model fragments' inherent dichotomy of totality and fraction enables the resultant performance in the staged environments of exhibition and theatre. The architectural model-fragment manifests as a scenographic and performative object that demands what it negates, namely completion, thus offering a spatially articulated criticality. The architectural model-fragment's 'chaotic universality', to cite Schlegel again, stirs up and questions any unified representation. This discourse is one of interruption and intentionality that invariably must speak of the representation of memory, identity, absence and presence as motifs embedded into models of an incomplete architecture. Reversing German idealist philosopher Georg Wilhelm Friedrich Hegel's dictum 'The truth is the whole', Theodor W Adorno famously posited 'The whole is the false',[4] thus articulating his mistrust towards the perfection of completion.

The research for this article was informed by two central works that approach the fragment in a manner not unlike Adorno's, in terms of their scepticism toward the whole. Here, it is argued that the reader's and the viewer's capacity to desire the whole (world) in the model-fragment lies in the unchanging nature of the

The architectural model-fragment in its manifestation on the stage and in exhibition is a novel phenomenon of our time in that it performs the postdramatic maxim of the deliberately unfinished, the made and the constructed

Heimo Zobernig,
Austrian Pavilion,
56th International Art Biennale,
Venice,
2015

Installation view showing the void left from the removal of the pavilion's back wall, opening the internal space out into the exterior garden and rendering the building incomplete.

fragment rather than in history. The contemporary viewer is neither more experienced nor more willing to enter into the process of worldmodelling than an 18th-century reader, but rather the architectural model-fragment in its manifestation on the stage and in exhibition is a novel phenomenon of our time in that it performs the postdramatic maxim of the deliberately unfinished, the made and the constructed.

Comprising contemporary architectural model-fragments on the theatre stage and in exhibition respectively, the investigated works here are German scenographer Anna Viebrock's set design for the premiere in 2019 of René Pollesch's *Life On Earth Can Be Sweet (Donna)* at the Deutsches Theater Berlin, and Austrian artist Heimo Zobernig's 2015 untitled intervention for the 56th International Art Biennale in Venice for the Austrian Pavilion. In creating staged architectures, Zobernig and Viebrock explore the performative quality of temporary architectural model-fragments in the live situations of theatre and exhibition respectively. Citation, as well as self-referentiality, are the main strategies discussed here; that is, to ask what the model-fragment *does* within a staged and altered architectural exhibition and what it *does* on the theatre stage. These projects, conceived by non-architects, guide our discussion on the architectural fragment beyond its more frequent appearance as *spolia* or some form of intentionally built ruin. Viebrock's and Zobernig's works offer, through the performative, amongst other strategies, a critical voice on architecture utilising the worldmodelling and worldmaking capacity of the architectural model.

Viebrock is known for architectural scenographies collaged from photographs of existing, dire 1960s interiors. Meticulously reproduced, researched and assembled, they typically tend to include surreal elements added by Viebrock to subvert the functionality, scale or perspective of an otherwise unremarkable, ordinary and often oppressive hotel lobby, office or apartment block. Working across the genres of theatre, dance, opera and (model) exhibition, she holds the chair for scenography at the Academy of Fine Arts Vienna where, coincidently, visual artist Heimo Zobernig holds the chair for sculpture.

Zobernig's recent architectural interventions are informed by a past practice that has operated across a diverse palette of media from painting to video, pixels to monochromes and a desire for the void or emptiness that coalesce in Venice.[5] As he remarks, 'In the early Eighties I created a lot of small maquettes of architecture which I understood as small sculptures.'[6] Educated as a set designer, his focus on the relationships between object, audience and the spaces they inhabit has been a consistent theme. His body of work resonates in the architectural intervention seen in Venice, and recent, related projects in Bregenz, Malmö, the Massachusetts Institute of Technology (MIT) and Vienna. In a subtle critique of modernism, Zobernig transforms existing spaces through architectural interventions that often do not announce themselves but rely on the visitor's presence to reposition the artist, institution and artwork.

A Performance of the Architectural Model-Fragment

For *Life On Earth Can Be Sweet (Donna)*, Viebrock brings onto the revolving stage of the Deutsches Theater a collection of walls of different heights as well as surfaces and a glassed-in shop window. Disturbingly unrelated to each other and negating conventional parallel grid-plan orientation, this configuration of architectural fragments forms an incomplete, improbable architecture of anonymous urban interiors and laneways. The walls, built as typical theatre flats, are supported by scaffolding visible on the reverse side. In laying open the construction techniques of theatre architecture on the stage by revealing the usually unseen scaffolding, Viebrock presents an anti-illusionist scenography that spatially mirrors Pollesch's labyrinthine-discursive text and its explicit model character.

Five actors playfully explore different modes of representation concerning emotion and identity loosely based on playwright Bertolt Brecht's basic model of 'epic theatre' with his example of the 'street scene'.[7] An eyewitness describes an accident to bystanders in a demonstrative (objective) rather than an emotive (subjective) way. Analogous, theatre's political agency is only achieved by the actor as a demonstrator with the stage as a functional rather than aesthetic environment. Viebrock and Pollesch translate Brecht's didactic street scene spatially and textually into a whirlwind-like self-referential treatise on the theatre itself, performed by actors who seem to meet each other by chance in the laneways complete with cut-out cardboard cars 'driven' by the performers: 'Yes and now it is important to inform everybody. To drive around with the car and tell it to the people.'[8] Here, Brecht's theoretical theatre model is articulated as the physical manifestation of an idea of the theatre today – as a model. Viebrock cites Brecht's realist setting of street, lane and corner but treats each element separately, resulting in an incomplete stage architecture set upon a relentlessly stuttering revolving stage.

Scenographer Anna Viebrock transforms the proscenium stage of Berlin's Deutsches Theater into a disorienting architectural fragment. The photograph, taken from the production desk during a rehearsal, shows the stage in its full depth that reaches to the existing back wall of the stage.

Notions of 'inside' and 'outside' are deliberately blurred in Viebrock's streetscape that reveals both front and back of the theatre flat, thus creating an anti-illusionist scenography.

```
Anna Viebrock,
Sets for René Pollesch's play
Life On Earth Can Be Sweet (Donna),
Deutsches Theater,
Berlin,
2019
```

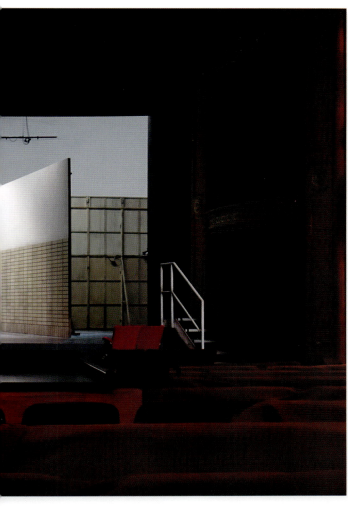

Populated by actors questioning who they are, which acting style they should adopt and where they should position themselves, the revolving stage with its intentionally unreliable mechanism lays bare the theatrical contract between actors, stage and audience to be a fragile construct. In rapid-fire-like, dizzying and cascading monologues, the five actors articulate the uncertainty of human existence through the lens of the performer trapped in the labyrinth of the architectural model-fragment.

Addressing the audience directly in a gesture Brecht stipulated as one of the central elements of 'epic' in contrast to 'dramatic' theatre, the performers while being whirled about by the revolving stage, confirm theatre's agency to challenge societal norms discursively. The performative world of the postdramatic theatre,[9] with its stubborn central questions of the 'hows' of the representation of emotion, identity, body, text and space, exists scaled down to the dimensions of the Berlin stage.

Viebrock's scenography accompanies and underscores these 'how' questions by creating a model of an intentionally incomplete (fragmented) architecture of and for a space that oscillates between the uncertainty of the 'real' and the possibility to reinvent the 'real' in performance. In Viebrock's architectural model-fragment, representation is nothing more than a discursive proposition that is open for negotiation. Here, the unique performance of the architectural model-fragment comes into play in that it has inscribed into its intentional incomplete presence the absence of the whole.

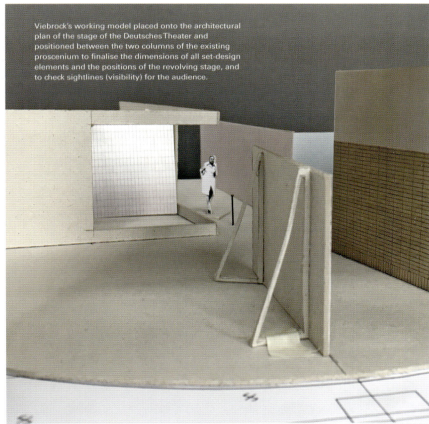

Viebrock's working model placed onto the architectural plan of the stage of the Deutsches Theater and positioned between the two columns of the existing proscenium to finalise the dimensions of all set-design elements and the positions of the revolving stage, and to check sightlines (visibility) for the audience.

The Fragment in an Infinite Field

While Viebrock designed and built an intentionally incomplete theatre architecture inhabited by actors, Zobernig, in Venice, turned a complete architectural object into an incomplete (and ruinous) model-fragment inhabited and restored by the visitor. His 2015 intervention operates in and with the Austrian Pavilion in the Biennale's Giardini, designed by Josef Hoffmann and Robert Kramreiter in 1934 and extensively restored by Hans Hollein in 1984. In its original state, the pavilion is dominated by a strict symmetry and mirror-reversed primary and secondary spaces across a central access. The intent of its clear, rational modernist spaces and materials appears to be repudiated by the more eclectic classical arches and centralised hierarchy that Hoffmann is better known for, alternating the stylistic reading of the building from Italian rationalism to modernism and historicism.

Five discrete spaces of differing heights make up the pavilion: the central corridor leading from the entrance to the garden across the breadth of the building; the two main exhibition spaces; and two secondary exhibition spaces. Zobernig's immersive intervention into the Austrian Pavilion, as many of his other works, is not immediately discernable to the casual viewer who needs to rely on a previous visit, catalogue text or understanding of the spatial language of Hoffmann or Zobernig to differentiate the intervention from the building. The artist subverts Hoffmann's hierarchy of spaces (the visitor may remember the compression of spaces as they moved away from the central axis in previous visits), by levelling the floor and ceilings. Through the inclusion of a black floor and monolithic ceiling, a uniform floor-to-ceiling height is enforced to 3.6 metres (11 feet 10 inches), thus negating the monumentality and verticality of the existing central interior and creating a rectilinear horizontally orientated volume that opens out to the rear.

This volume intentionally allows the visitor to enter the work not only experientially and intellectually but as an integrated compositional element in the artwork 'much like a figure in a painting's relationship to its background'.[10] The suspended ceiling masks Hoffmann's arches, skylights, windows and other 'historicising architectural elements',[11] creating a compressed black horizontal void interrupted only by the white walls and columns of the pavilion. Zobernig further subverts the strict parallel geometry of the central axis by removing the pavilion's back wall, allowing for an uninterrupted flow from architecture to garden that is an intentional remnant of the 2014 exhibit for the Architecture Biennale designed by landscape architects Auböck + Kárász. Zobernig's Venice intervention argued here as a material fragment in an imagined continuum leaves the possibility for further fragments to be placed at other times or locations in the future, as he did at the Kunsthaus Bregenz the following year. Zobernig's intervention is 'both site-specific and an autonomous object, which can and will later be shown in other contexts'.[12] Hoffmann's architectural object is transformed into an architectural fragment within a historical and spatial continuum 'that continually travels and thereby refers to its past, present and future places'[13] while simultaneously adopting model characteristics that allow an immaterial world to be imagined by visitor and artist alike.

The artist demonstrates that the interventionist act disturbs the pavilion's discrete object character, yet the intervention has been merged with the original to erase some of its features. The removal of the back wall of the building renders it a ruin open to the elements and incapable of functioning as originally intended. Consequently, there are no exhibits, allowing Zobernig's model-fragment to speak to the architecture itself, its potentiality to construct and intervene into and shift its fragile composition between material, visitor and atmospheric presence. Ultimately, Zobernig creates a work 'where one can linger and reflect on human presence in space'.[14] The horizontal intervention infers an extension beyond the limits of its

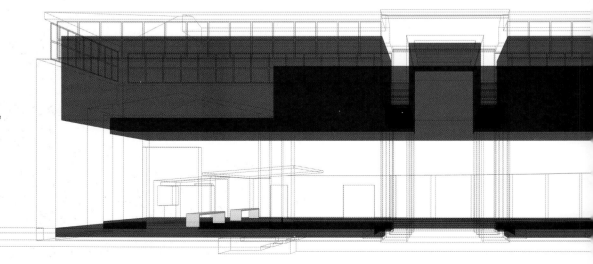

Heimo Zobernig,
Austrian Pavilion,
56th International
Art Biennale,
Venice,
2015

CAD visualisation of the black monolith structure and black floor used by Zobernig to create a single rectilinear volume from Josef Hoffmann's original assemblage of volumes.

Exterior view of the pavilion, designed in 1934 by Josef Hoffmann and Robert Kramreiter, showing the view through the building to the garden beyond and a glimpse of Zobernig's black monolithic intervention.

Interior view showing Zobernig's intervention into the pavilion's central axis. His response transforms the original arches into columns and excludes the vertical monumentality and formal hierarchy of the original interior.

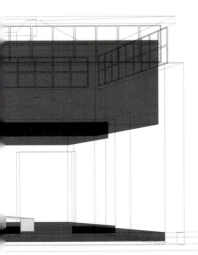

materiality, continuing outwards in a way reminiscent of the graphics and propositions of the 1960s Italian Radicals' notion of architecture as an endless field or grid and recollective of San Francisco Museum of Modern Art curator Joseph Becker's provocation 'What if a wall or a floor went on forever?'[15] from his 2012 exhibition 'Field Conditions'.

The Non-hermetic Architectural Fragment
Unlike the architectural plan or drawing, the model is uniquely positioned to create whole worlds. Its cosmopoietic capacity becomes even more compelling when the model itself comprises but a fragment. Artists Viebrock and Zobernig produce non-hermetic architectural fragments by employing walls and ceilings as delineating elements. These fragments are not remnants of the past, but rather manifest as intentional constructions of incomplete architectures that comment discursively on the present. As enterable models, their materiality and geometries extend into the immaterial; the visitor desires to complete the composition and becomes complicit to the worldmodelling. The fragment solicits visitor/performer participation in its completion, resulting in each imaginary world being both distinctive and simultaneous. The architectural model-fragment operates as the material representation of a real or conceptual remnant of a shattered mass or exists as the origin of a world and infers a performative, charged and heightened notion of the whole to be a futile construct. ∅

Notes
1. On cosmopoiesis, see Thea Brejzek and Lawrence Wallen, *The Model as Performance: Staging Space in Theatre and Architecture*, Bloomsbury (London), 2018.
2. Friedrich Schlegel, *Athenäums-Fragmente* [1789], Edition Holzinger (Berlin), 2016, p 118. Translation from the German original by Thea Brejzek.
3. *Ibid*.
4. Theodor W Adorno, *Gesammelte Schriften*, vol 4: *Minima Moralia, Reflexionen aus dem beschädigten Leben: Zwergobst*, Digitale Bibliothek Band 97, Suhrkamp Verlag (Frankfurt), 1980, p 1726. Translation from the German original by Thea Brejzek.
5. 'Heimo Zobernig in conversation with Mark Godfrey (Tate Modern)', https://vimeo.com/212229152.
6. *Ibid*.
7. Bertolt Brecht, 'The Street Scene: A Basic Model for an Epic Theatre' [1950], in John Willett (ed and trans), *Brecht on Theatre: The Development of an Aesthetic*, Methuen (London), 1964, pp 121–9.
8. René Pollesch, 'Lied vom Ungebundenheitsimperativ': www.deutschestheater.de/programm/a-z/life-on-earth-can-be-sweet-donna/. Translation from the German original by Thea Brejzek.
9. See Hans-Thies Lehmann, *Postdramatic Theatre*, Routledge (London), 2016.
10. 'The Venice Questionnaire 2015#17: Heimo Zobernig', *ArtReview* online exclusive, 28 April 2015, https://artreview.com/2015-venice-17-heimo-zobernig.
11. 'Heimo Zobernig in conversation with Mark Godfrey (Tate Modern)': https://vimeo.com/212229152.
12. 'Exhibition Heimo Zobernig', 2015: www.phileasprojects.org/2015-exhibition-heimo-zobernig.html.
13. *Ibid*.
14. Heimo Zobernig, *Untitled*, Biennale Arte: The 56th International Art Exhibition, Austrian Pavilion, Venice, 2015.
15. Joseph Becker, 'Field Conditions', 30 October 2012: https://openspace.sfmoma.org/2012/10/field-conditions-i.

Text © 2021 John Wiley & Sons Ltd. Images: pp 74–5, 78–9 © Anna Viebrock; pp 76–7, 81 © Photos Georg Petermichl / Archiv HZ; pp 80–1(b) © Eric Kläring / Archiv HZ

MODELS AS OBJECTS

THE INSTALLATION AS ARCHITECTURAL ENCOUNTER

James A Craig and Matt Ozga-Lawn

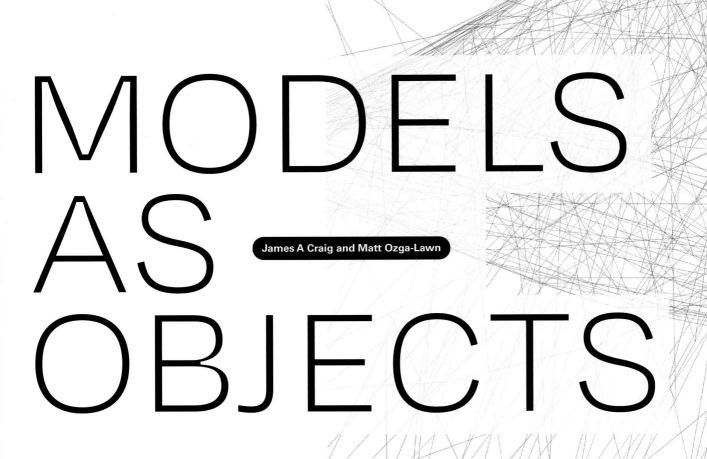

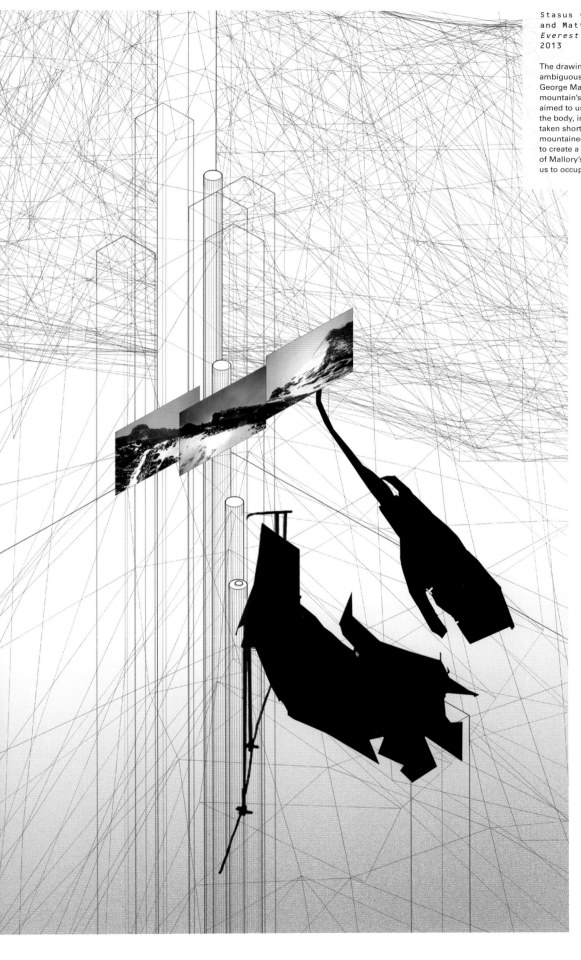

Stasus (James A Craig and Matt Ozga-Lawn), *Everest Death Zone*, 2013

The drawing depicts the ambiguous relationship between George Mallory's body and the mountain's summit. The project aimed to use evidence from the body, including photographs taken shortly before the mountaineer's disappearance, to create a spatial understanding of Mallory's death and so allow us to occupy this moment.

Transgressing traditional conceptions of the model and architectural intervention, in the work of their practice Stasus **James A Craig and Matt Ozga-Lawn** seek to create a sense of suspended pleasure by partially accommodating the viewer in their installations, giving rise to fleeting fluctuations in the perception of the architecture and its site history.

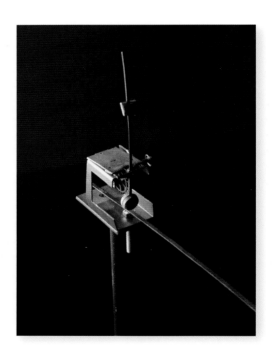

Stasus (James A Craig and Matt Ozga-Lawn), *Animate Landscapes*, 2009

right: The reconstituted metronome. This symbolic object structured the project through the manifestation of a rhythm dictating the accretion of further objects and accompanying architectural elements. It also served to set the project apart from its surroundings in Warsaw as if occurring with a different sense of time passing, so preserving the site against the rapidly encroaching city centre.

below: Drawing showing the reconstituted metronome, now transformed to a scale of 1:50 and imagined as a dwelling for a timekeeper – a custodian of the landscape.

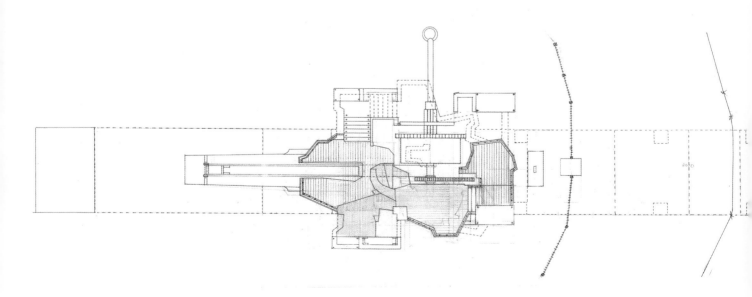

The architectural model has been comparatively under-studied in relation to other key components of architectural representation, particularly the drawing. In post-Structuralist discourse that architecture involved itself with in the 1970s and 1980s, the drawing became the pre-eminent mode of dealing with complex, often abstract ideas in discontinuous methods, allowing for the 'generative performance' of architectural spaces in the imagination.[1] This necessitated the abandonment of the figurative and literal, which the model was generally seen as allied to, and the resulting 'paper architecture' conveys its drawing-based nature upfront.

This is unfortunate, as the model has the capacity to act in unexpected and compelling ways when considered primarily as an object. The nature of objects and our relationships with them as they are transformed and manipulated is instrumental in contemporary art, particularly installation art which has, as yet, had little influence on modes of architectural representation. By considering architectural models as objects, they are opened to a rich theoretical discourse, including object/subject dialectics that further elucidate issues around the reading of architectural models and drawing. Finally, the model-as-object allows us to better consider our embodied practices around the reading and production of architectural ideas. Two of Stasus's projects utilise the nature of familiar objects and their deterritorialisation to create complex, performative and imaginative architectures through partial worldmodelling, subjective interpretation and embodied narratives.

The Metronomic Landscape

The Animate Landscapes project (2009) started from the seeking-out of an object that could embody the atmospheric conditions of a site in western Warsaw. The site holds a troubled history, including massacres and the clearance of domestic space. To us it was like a petrified island, a near-empty peri-urban wilderness set against the encroaching backdrop of Warsaw's central business district. We wanted to symbolise the stillness and impending threat through a found object that would have the capacity to be read at multiple scales – to be held in the hand at 1:1 and to be imagined on the site at 1:100. The found object was to function as a vessel to hold our projected perspectives on the landscape, acting as a kind of souvenir of the site.

The reconstituted metronome signalled the first in a series of found, principally domestic objects that would come to symbolise aspects of the Warsaw site. Each found object is reconstituted into a series of architectural models that look at the landscape in different ways, and that are inspired from the palimpsest of moving bodies that once occupied the site. Each of these objects is held in tension in an installation, demanding to be noticed by the viewer in its capacity to physically engage with a body. In this way, the installation becomes a space that holds onto the models, but also allows the viewers to look at their own subjective engagement with the site. Here there is a mixing of conditions that destabilises the viewing subjects; they are invited to perform the architecture by moving through the installation, and their movements become symbolic of the actions that have been designed in the proposal at multiple scales. To walk around the installation is, in a way, to walk around the site.

Photograph showing the bodily engagement with the installation. The viewing apparatus works as a navigational tool to look at the site at multiple scales.

The Bodies of Everest

The Everest Death Zone project (2013) comprised mappings, objects and an installation reflecting on the first summit attempt of Mount Everest in 1924, which was also the final ascent of George Mallory. Where Animate Landscapes started with an object in the metronome, Everest Death Zone starts with a body. Both projects develop a landscape from these starting points, emanating outwards from the initial object, until an embodied experience of the accrued models, drawings and their projected spaces can be encountered through installation. Inspired by the International Necronautical Society's manifesto,[2] in which they outline as one of their supposed aims the mapping, colonising and inhabiting of the space of death, we used Mallory's ascent/ascension as the locus of a study of a landscape comprised of death: the Everest Death Zone is the terrain of the mountain 8,000 metres (26,250 feet) above sea level, in which a human cannot survive unaided for long, with more than 150 bodies remaining unrecovered in the area.

The drawings and installation spatialised the circumstances of Mallory's death through a reading of his body, found preserved intact some 70 years later. Objects, such as his climbing axe and mountaineering rope, were reconfigured to be at the centre of a performative installation that also utilised footage and

Stasus (James A Craig and Matt Ozga-Lawn), *Everest Death Zone*, North Tower, Tyne Bridge, Newcastle-upon-Tyne, England, 2016

right: The installation deployed mountaineering equipment such as guide ropes and carabiners suspended in the North Tower. The vertiginous void in the tower recreates (to a limited extent) the relationship between the mountain surface and its fragile occupants, helping us interpret their experiences.

below: Projection of the film *The Epic of Everest* onto the installation fabric. The film was the official record of the 1924 expedition produced by Captain John Noel and recently restored by the British Film Institute. As Mallory's body was not found until the 1990s, the film ends with an ambiguous sense of the summit attempt as a noble death.

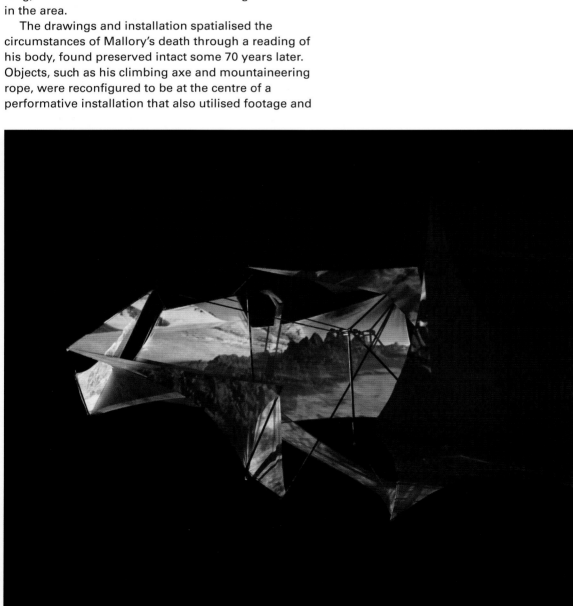

photography from the summit attempt.[3] The installation, initially constructed in the cavernous void within the Tyne Bridge North Tower in Newcastle-upon-Tyne, England, as part of the 'Being Human' festival of the humanities,[4] and then digitally reconstructed as a virtual-reality environment, orients us through the body of Mallory, viewing towards the distant summit of the mountain. It has never been ascertained whether Mallory (and his climbing partner) died ascending or descending from the summit, and so we are left unsure as to whether the summit attempt was successful or not. We use this ambiguous relationship between his body and the summit to stage the work. The body becomes an object, itself reconstituted and redesigned with mountaineering elements into a kind of shelter through which to observe the mountain's iconic summit, and so to reflect on the relationship of the elements depicted to the observer's experiencing body. The immediacy provided by the figurative elements allows a different reading of the more abstracted elements, a transgressive bridging of radically different representational frameworks made possible through the installation environment and the objects contained by it.

Sites of Encounter

Stasus's 'architectural installations' work as intermediary spaces that facilitate embodied encounters with architectural representation, actively resisting the distancing effects that stem from perspectival drawing modes and the miniaturisation of traditional forms of architectural modelling. The installation sets up a performative playing-out of the architecture in the imagination of the observer. This act of imaginative configuration has been termed 'suspended pleasure',[5] alluding to the fact that what is really occurring is a discontinuous jumping between radically different, potentially even incompatible representational modes. By building on this discontinuity, and foregrounding it, architectural installations might operate closer to the work of installation artists like Bruce Nauman and Dan Graham, in which the viewers are accommodated, but only partially, creating a situational unease (and self-consciousness) about their role as observers: becoming both subjects and objects of the work. Architectural drawings rarely ask their audiences to reflect on their bodily relationship to them, opting instead for a cinematic immersion within, or an abstracted distancing from, their heavily coded phenomena. By refocusing on the observer's interactions with drawings and models, architectural installations frame our engagement with them as a site of encounter. ⌂

Stasus's 'architectural installations' work as intermediary spaces that facilitate embodied encounters with architectural representation

Notes
1. Robin Evans, 'In Front of Lines That Leave Nothing Behind', *AA Files* 6, 1984, pp 89–96.
2. www.necronauts.org/.
3. *The Epic of Everest*, director John Baptist Lucius Noel, 1925.
4. https://beinghumanfestival.org/.
5. Penelope Haralambidou, *Marcel Duchamp and the Architecture of Desire*, Ashgate (London), 2013, p 12.

Text © 2021 John Wiley & Sons Ltd. Images © Stasus

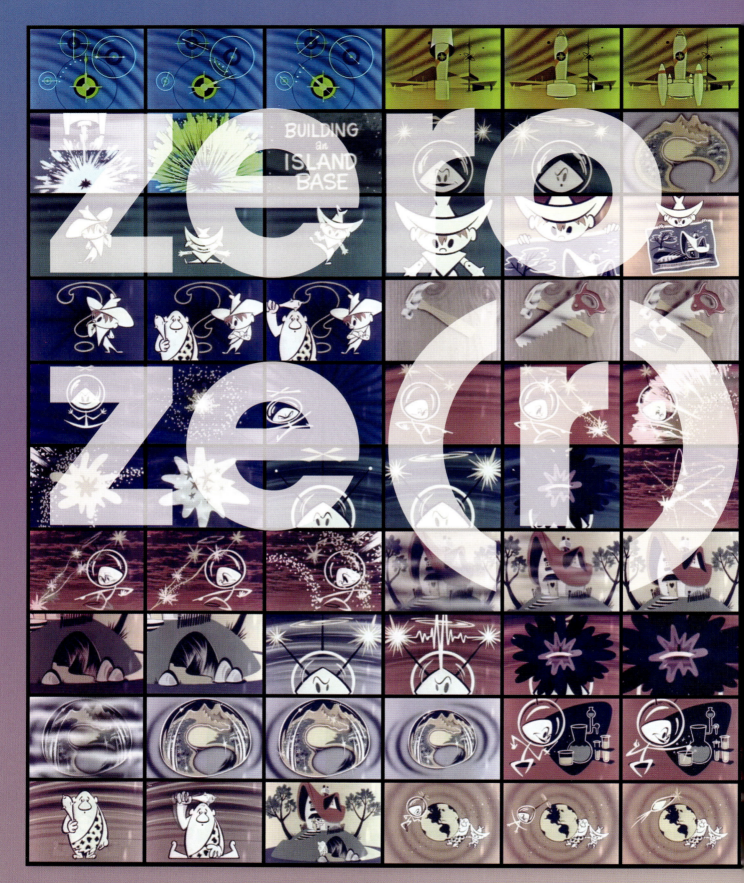

Robert D Buchanan,
'Building an Island Base' episode,
Colonel Bleep,
Soundac, Inc,
1957

The sequence of stills is from the first colour cartoon on television, in which the main characters Colonel Bleep, Squeak and Scratch arrive at their new home base at Zero Zero Island, and the blueprint is the starting point for them to envision a new world that will include a house, cave and laboratory underneath a dome for protection from enemies. For every built world a mapping is required not only to navigate the terrain, but also to construct it. Compilation of images by Ryan Dillon.

Ryan Dillon

zero to (r)

How the Cartographic Thirst to Project the Real Reveals Spaces for the Creation of New Worlds

Through a cartographic dérive of Greenwich, home of the Prime Meridian, Architectural Association Head of Communications Ryan Dillon reveals lines, variances, elisions, paradoxes and lies in a world that cannot be totally known or described. His trajectory provokes musings on the inconsistencies of the map and the modelling of any terrain.

A map is the greatest of all epic poems. Its lines and colours show the realisation of great dreams.
— Gilbert H Grosvenor[1]

A map. A town. A paper. A street. A cartoon. A griffin. A meridian. In Greenwich. Let's try to follow a line. Of thought. Of time. Around the world. On a crisp October afternoon, I stand within the shadow of General Wolfe, the statue, not the man. Inputting coordinates into my mobile phone, the Google Maps blue line instructs me to head down the hill due east and past the sundial in which the gnomon casts a shadow onto the horizontal stone surface that tells me it's 12 pm. It's 11.50.[2] A gap. A trap. An error.

Located within the inconstancies of navigation and time, is space in which worlds can be built – otherwise known as *worldbuilding*. Cartography, a key component in the construction of new worlds, has historically attempted to make sense of the world. However, maps of planet Earth are a chimera; an unattainable quest for complete projected accuracy of a sphere at a celestial scale onto a two-dimensional surface at a table-, book- or pocket-scale. And yet, this did not stop the likes of Gerardus Mercator, James Gall and later Arno Peters or R Buckminster Fuller to think their methods would crack the code and represent the world in all its accurate precision. This is of course a fallacy. As Ortelius is quoted in his attempt, the 1570 *Theatre of the World* atlas, 'in some places, at our discretion, where we thought good, we altered some things, some things we have put out, and others where, it if seemed to be necessary, we have put in'.[3] Located in the impossibility of mapmaking, and the very subjective nature of it, emerges the creation of a fiction, much like a novelist who creates a story – a world in which the reader can enter – from the reality that surrounds us.

The Transformative Nature of Lines
As I continue my journey, attempting to walk a line, I reach the edge of the River Thames, a key artery to British navigation, and the catalyst for an empire that relied on cartography to traverse the globe. None of this would be possible without lines. Dating back to AD 70–130, Marinus assigned two types of lines – latitude and longitude – in an effort to demarcate every location he aimed to map. As Dava Sobel correctly points out, while the location of the equator and all other 'parallels' are fixed by nature, the placement of a longitude, especially 0° of the world, is inherently political[4] – and anything political is a construct of human invention. Therefore, these lines begin to splinter into the realms of both politics (as a means to control the world[5]) and fiction (an alternative to this control); both can be understood as *worldbuilding*.

To follow a line that crosses through three continents, nine countries, four seas, three oceans, a gulf, a channel and numerous lines, many myths and worlds begin to emerge, starting with King Neptune – the god of the sea in Roman religion. Charles Darwin, referring to himself as a 'poor *griffin*', reminisces in his 17 February

1832 *Beagle Diary* entry entitled 'Equator' of being blindfolded, doused with buckets of water and led to a plank, and describes the scene as the 'whole ship was a shower bath: & water was flying in every direction: of course not one person, even the Captain, got clear of being wet'.[6] The start of this diary note was, 'We have crossed the Equator, & I have undergone the disagreeable operation of being shaved.'[7] Here, Darwin is referring to the ritual of a line-crossing ceremony in which those on board a naval ship worship King Neptune as the rank of 'griffins' and are transformed into 'trusty shellbacks', albeit hairless as they cross the line of importance. Those who are willing to go through this hazing more than once can gain the sparkling designation of Royal Diamond Shellback[8] if they cross the point where the equator and Prime Meridian intersect. To document this momentous occasion a photo-op presents itself within the Gulf of Guinea where nothing but the vast sea and single weather buoy marking 0° N, 0° E sits in the background.[9]

Ezgi Terzioglu,
Crossing the Line,
Intermediate 5 unit,
Architectural Association,
London,
2018

left: The map by student Ezgi Terzioglu is of the International Date Line, a non-legal outcome of the Greenwich Prime Meridian, which is transformed into a fluid, constantly changing line that takes its shape from data obtained from the average locations of fishing activity in the Pacific Ocean. The outcome of this fluidity manifests itself in a world in which vessels crossing the line can time travel, extending fishing hours for the political and economic benefit of the island nations.

above: The vessels, which reference lighthouses, light ships and the physical objects that mark the Greenwich Prime Meridian, move in parallel to and cross the fluid International Date Line. This condition creates a non-time territory where the inhabited architectures, occupied by fishermen in the Pacific Ocean, legally drift along two calendar dates, resulting in a space of ambiguity that breaks the perception of spatial orientation and the flow of time.

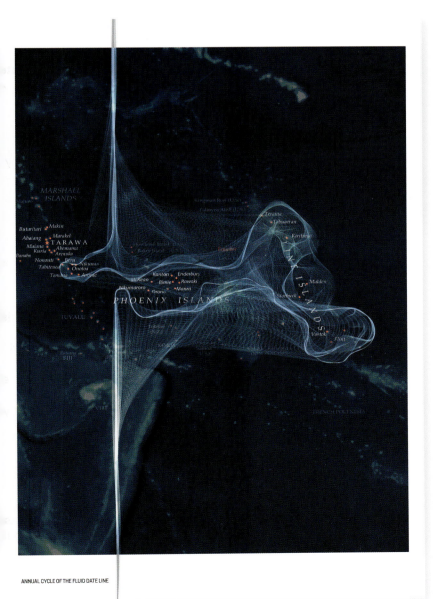

ANNUAL CYCLE OF THE FLUID DATE LINE

An Island and Many Worlds

Now close your eyes, listen to the rhythm of the waves and drift off into another world. Welcome to Zero Zero Island. Home of Colonel Bleep and his team, Scratch and Squeak. In 1957 Richard D Buchanan used this location for his animated series, the first colour cartoon in television history, and to build a world. The opening line for each of the 100 episodes starts with, 'And now, stand by for adventure …' and in one of the earliest of these, entitled 'Building an Island Base', the three characters touch down onto a crescent-shaped landmass with mountains to the north, forest to the east and west and a flat open beach-like *tabula rasa* anchoring the central area that is connected to what looks like a natural port for passing-by ships to dock. Squeak, who represents the present, draws a blueprint of his dream house and Colonel Bleep, with his futuristic wizardry, zaps it into realisation. The home sits on a large rock formation, which acts as both foundation and a cave, carved out for Scratch, who as a caveman is an agent of the

Sinan Asdar,
A Domestic Glitch,
Intermediate 5 unit,
Architectural Association,
London,
2019

The street depicted in this plan becomes a motherboard for a new type of kinetic and decentralised housing that is in flux with spaces being erected and dismantled over time. Student Sinan Asdar's project is sited on Whitfield Street, London, a cartographic trap, allowing a new world to form in which data collected from inhabitants is used to generate an algorithm that reorganises the architecture on an as-needed basis. An address with postcode is replaced by an IP address.

Sofia Belenky,
Null Island,
Intermediate 5 unit,
Architectural Association,
London,
2017

left: The map by student Sofia Belenky is of Null Island, a data wasteland where a series of sandbars expand and contract depending on the time of day and that is dependent on the tides for occupation within the pixelated sea. The structures, as seen in the map insets, include a welcome centre, resort, post office and data centre, all of which are connected by bridges that symbolise the spirit of union and participation in a post-truth world driven by broken data.

above: The drawing is of the Data Center at Null Island, which houses information with errors in its geocoding sent to the 0,0,0 by geographic information system (GIS) analysts. The interior of the building is not accessible but contains hundreds of servers protected by a system of mesh layering that additionally provides protection against data leaks. The screens located on the façades display missing pixels and the mirrors reflect the surrounding context of ocean as a means to camouflage the structure.

> The Republic of Null Island, a new 'country', is two islands: a virtual one and a fictional one, both constructed by and for error

past. Reminiscent of Fuller and Shoji Sadao's proposed geodesic structure over Manhattan (1960), a 'gigantic plastic-like dome' covering all of Zero Zero Island is erected to protect from 'bad weather and insects' and serves as a laboratory for Bleep's experiments and research. Part science fiction and part educational, the Colonel Bleep series is an early example of using cartography and the lines required to map for the location of a new world and to 'establish contact with the real experimentally'.[10]

This overlap of fictional and real can be seen in Zero Zero Island's alias: Null Island. A different world in the same location. Several thousand years BC, a volcano erupted beneath the Gulf of Guinea, approximately 350 miles (560 km) due south from the southernmost tip of Ghana. As of 2011 a fully formed island at this position was listed on Wikipedia. A beautiful landscape, full of hot springs, crater lakes and lava domes, and surrounded by clear waters full of marlin, mako sharks and swordfish, became an ideal – fictitious – holiday spot. Geographers who discovered the landmass were quick to act, coined its name and began a campaign to anoint this the world's newest country. First came the map, more graphic than nautical; followed by a flag with a grey circle reminiscent of a telescope lens with cross hairs (or two lines); the development of a language (Nullian); and the naming of inhabitants (Nullians). Despite warnings from scientists to not visit due to an active volcano that could erupt at any moment, this small island became the most visited place on Earth. Yet, no human has ever set foot on Null Island. If, as Gilles Deleuze states, the act of writing is about 'land surveying and cartography, including the mapping of countries yet to come'[11] then Null Island fits the bill. Coincidently, or not, in 2011 the map dataset developer Natural Earth entered 0,0,0 as coordinates into the public domain with the aim to help geographic information system (GIS) analysts to deal with errors in geocoding. Each time someone inputs a piece of text or

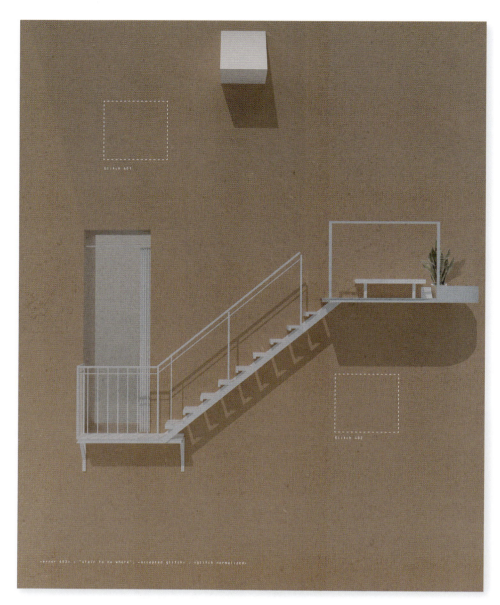

Sinan Asdar,
A Domestic Glitch,
Intermediate 5 unit,
Architectural Association,
London,
2019

Error 402: Stair to Nowhere. The project is generated through the collection of data and aims to create moments of social interaction – some random and enjoyable, others awkward and unsettling – that challenge the routine of everyday life controlled by international time zones. Error is seen to permit invention and intervention, creating an environment that embraces the glitch as a productive tool, resulting in unexpected and innovative visions for a new way of living in the city.

What seems to be an error can be seen by worldbuilders as simply an opening. A redirection to create a world; visions that make us think differently about reality

grouping of numbers, such as an address or the name of a place into a website, geocoding translates this into a set of coordinates. However, humans input the original information, and humans are error-prone. Due to typos or computational glitches a percentage of the data turns up as invalid, resulting in an inability to tag the material and archive it. These data mistakes end up at the virtual Null Island. Cleverly named 'null', drawing on both aspects of its definition (meaning both zero and invalid), The Republic of Null Island, a new 'country', is two islands: a virtual one and a fictional one, both constructed by and for error.

Two Lines Splinter into a Possible World …

As I wake up and return to Greenwich, still trying to follow the line that seems to elude me, I walk down General Wolfe Road, cross Shooters Hill and search for Whitfield Street. I check my compass, which situates me at 51°28'12.4"N 0°,0', 05.3101" W. I should be on 0°,0', 0" W, and I cannot seem to find Whitfield Street at all. There appears to be a misalignment between the old, tattered analogue A–Z map I still use from the 1990s that clearly demarcates a road and Google Maps that tells me no road exists. Reality confirms Google. Plagiarism in cartography, like writing, is a real thing. However, unlike writers, mapmakers have deployed visual techniques to thwart would-be copiers. A paper town or the implementation of a trap street, invented by mapmakers as fictional places, aim to 'trap' devious plagiarisers. Whitfield Street, perfectly aligned with the Prime Meridian, is a trap. However, what seems to be an error can be seen by worldbuilders as simply an opening. A redirection to create a world; visions that make us think differently about reality.

Concluding my journey, I re-enter Greenwich Park. The home of the Prime Meridian as it divides the architecture of the Royal Observatory, designed by Sir Christopher Wren. Like many tourists I straddle the line embedded in the ground that marks 0° longitude of the world to take a selfie. Looking at the geotag of the image, it provides 0° 00' 05.3101" as the line of longitude. It should be 0°,0', 0". I begin to drift east, determined to find the Prime Meridian once for all. In 1835, on his appointment as Astronomer Royal, George Biddle Airy looked to the sky, stars and planets to mark his place on planet Earth and ultimately developed the Airy Transit Circle, a telescope that determined 0° longitude. Airy, however, could have used some GPS – launch Google Maps at the Royal Observatory and you'll soon discover that his 1851 line is off by 104 metres (340 feet),[12] and the actual 0° is accidentally marked by a rubbish bin. I stand there now. Between two lines, one a fiction, one a reality, both arbitrary. A fissure – sounds like the perfect place to worldbuild … ↺

Notes

1. Grosvenor, the founding editor of *National Geographic* magazine, is credited with this statement: https://www.nationalgeographic.com/maps/about/.
2. The sundial in Greenwich Park had a series of issues such as a 3.5° error with the angle of the gnomon that causes inaccurate readings of time. For more information, see: www.thegreenwichmeridian.org/tgm/location.php?i_latitude=51.481332&i_type=|%20all%20markers%20|.
3. Abraham Ortelius, 'To the Courteous Reader', in Ortelius, *The Theatre of the Whole World*, English translation (London, 1606), unpaginated, quoted in Jerry Brotton, *A History of the World in Twelve Maps*, Penguin Books (London), 2013, p 10.
4. Dava Sobel, *Longitude: The True Story of a Lone Genius Who Solved the Greatest Scientific Problem of His Time*, Harper Perennial (London), 2011, p 4.
5. See Jonathan Martineau, *Time, Capitalism and Alienation: A Socio-Historical Inquiry into the Making of Modern Time*, Haymarket Books (Chicago, IL), 2015.
6. Richard Darwin Keynes (ed), *Charles Darwin's Beagle Diary*, Cambridge University Press (Cambridge), 1988, p 37.
7. *Ibid*.
8. Also called an Emerald Shellback. See Simon J Bronner, *Crossing the Line: Violence, Play, and Drama in Naval Equator Traditions*, Amsterdam University Press (Amsterdam), 2006.
9. National Oceanic and Atmospheric Administration's National Data Buoy Center, 'Station 13010 – Soul', https://www.ndbc.noaa.gov/station_page.php?station=13010.
10. Gilles Deleuze and Félix Guattari, *On the Line*, trans John Johnston, Semiotext(e) (New York), 1983, p 25.
11. Gilles Deleuze and Félix Guattari, *Anti-Oedipus: Capitalism and Schizophrenia*, trans Robert Hurley, Mark Seem and Helen R Lane, University of Minnesota Press (Minneapolis), 1983, p 247.
12. Stephen Malys, John H Seago, Nikolaos K Pavilis, P Kenneth Seidelmann and George H Kaplan, 'Why the Greenwich Meridian Moved', *Journal of Geodesy*, 89, 2015, pp 1263–72: https://doi.org/10.1007/s00190-015-0844-y.

Text © 2021 John Wiley & Sons Ltd. Images: pp 88–9 Courtesy of Ryan Dillon; pp 91–5 © Architectural Association School of Architecture

FROM M TO COU

SOME DIFFERENCES, CHALL OPPORTUNITIES OF BIO-HY

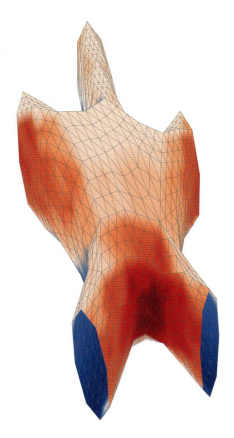
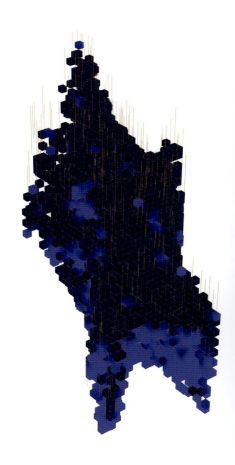

Adrien Rigobello,
Fungal Architectures simulation,
Centre for Information Technology
and Architecture (CITA),
Royal Danish Academy
School of Architecture,
Copenhagen,
Denmark,
2019

Current development of a modelling framework for simulating moisture distribution within mycelium composite components based on external environmental parameters and internal material structuring of the substrate. Moisture mapping is critical to simulating and monitoring the growth potential of the mycelium and correlates with changes in electrical resistance.

MIMICRY COUPLING

CHALLENGES AND BIO-HYBRID ARCHITECTURES

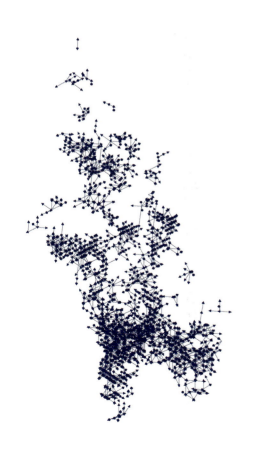

Phil Ayres is an associate professor at the Centre for Information Technology and Architecture (CITA) in Copenhagen. Here he discusses bio-hybrid structures and their novel applications in creating 'greener' worldmodelling architectural outcomes at odds with established formal tectonic tropes.

In 1959, Peter Collins, the British architectural historian and Professor of Architecture at McGill University, Montreal, published an article in the *Architectural Review* that traced the origins of scholarly use of biological analogy, to develop architectural theory and strategy, to the mid-18th century.[1] Since its publication, the realm of biology has continued to provide insight and inspiration within architectural design and production, but in ways that have deepened association from the predominately metaphoric and figurative towards worldbuilding models that privilege structural and performative characteristics. This enriching of association can be linked to the research field of biomimetics, which emerged and matured from the mid- to latter part of the 20th century. This field of study examines biological role models with the aim of transferring models and principles for application in artificial systems.

Coupling – an Alternative Worldbuilding Model
Mimicry can be a productive methodology, but it also has limits in the sense that it implies incorporation into existing approaches and methods of the application domain. Against a context of grand societal challenges in which the practices involved in the design and construction of the built environment have come under severe scrutiny, the question is raised as to whether transfers of understanding from biology into existing modes of practice can be consequential enough to elicit the scale of transformation necessary to balance degrading environmental effects. The need for alternative models is palpable.

Interestingly, we can find a more obscure approach being cultivated between the biological and the architectural from the start of the 18th century. This model can be defined through the notion of coupling – a more assertive and directly integrative approach whereby living biological complexes are conceptualised and incorporated as fundamental elements of the architectural proposition. Friedrich Küffner's *Architectura Viv-arboreo-neo-synemphyteutica*, published in 1716, provides an account of arboricultural practice directed towards architectural objectives.[2] Two centuries on, this fringe practice was still attracting scholarship. In 1926, the German landscape engineer Arthur Wiechula published *Wachsende Häuser aus lebenden Bäumen entstehend*, a text of practical and simple building techniques to be employed in the construction of houses made from living trees.[3] Almost a century further, this arboreal lineage of practice remains research active – a leading proponent being the Baubotanik group led by Ferdinand Ludwig.[4]

The field of bio-hybrid architecture extends this arboreal focus to widen the range of biological complexes targeted for architectural coupling. This represents a vast and underexplored territory, but one that is already being probed to investigate the roles that bacteria, mycelium and even social insects might play in the production, maintenance or decomposition of architecture.

Key Differences and Challenges with Bio-hybrids
Bio-hybrid architectures are defined through the symbiotic coupling of living biological complexes with artificial elements to create a dynamic system for achieving architectural objectives. The symbiotic coupling is what distinguishes bio-hybrids from more conventional approaches to the incorporation of biological elements, this being generally approached using a strategy of independent layering upon standard construction. The symbiotic coupling is also the locus of novelty and innovation in achieving architectural objectives, indeed, even creating the platform for the extension and invention of entirely new forms of architectural objective. The symbiotic coupling is the primary focus of study and design effort, and one that raises significant challenges to orthodox approaches of design.

For example, bio-hybrids exist in states of continual growth and adaptation, a condition that frustrates the conventional architectural aim of producing fully predetermined design targets that are realised through a discrete and finite phase of construction using a defined inventory of parts. They require methods for projecting and assessing their growth career in order to find and exploit architectural opportunity throughout their life cycle. This is necessary not only in the conventional 'design phase', but throughout the full life cycle of the architecture in order to regulate against the complexity of real-world conditions. This challenge points towards the need to expand orthodox modelling practices, models and interfaces to allow continual feedback throughout the life cycle of the architectural system, not only for assessment but as a control for feed-forward (projective) modelling that exploits the living, adaptive characteristics of the system and offers the possibility of modifying architectural objectives over time. This segues into the next point, that high-level architectural objectives for bio-hybrids require translation into methods for steering low-level self-organising behaviours of living specimens. This requires modelling methods that can 'compile' designs into instructions for relevant control mechanisms.

These three challenges identify the need for existing design approaches to be extended, enriched and revised to support a practice of continuity and persistence to enable effective engagement with the attributes of bio-hybrid architectures. They are also essentially engineering challenges, thus likely solvable. A fourth challenge is the more prickly issue of cultural conditioning and the indoctrination of expectations and values for architecture – what ideas it embodies, how we expect it should be, what we expect it to do, and how we expect it should (or should not) behave in doing it. Concrete proposition puts forward alternative values, providing material that can provoke the reshaping and shifting of entrenched and habituated expectations.

Bio-hybrid Opportunities Investigated Through Proposition

The argument that a bio-hybrid model for architecture produces qualitatively different architectural outcomes is supported with reference to four projects: Flora Robotica (2015–19), an EU-funded cross-disciplinary research project comprising a consortium of six partners from across Europe;[5] Co-occupied Boundaries (2016) by Asya Ilgun, produced within the Centre for Information Technology and Architecture's (CITA's) Computation in Architecture Master's programme at the Royal Danish Academy School of Architecture, Copenhagen, Denmark; Restless Labyrinth (2020) by Claudia Colmo, produced within the same Master's programme; and Fungal Architectures (2019–22), an ongoing EU-funded cross-disciplinary project comprising a consortium of four partners from across Europe.[6]

Each project examines different and novel architectural objectives that result from the specific biological coupling investigated. These are: (1) steered growth of material into structural configurations; (2) multispecies co-habitation; (3) functionalisation of bio-materials; and (4) spatial reconfiguration driven through site remediation processes.

The objective of the cross-disciplinary Flora Robotica research project was to develop and investigate symbiotic relationships between robots and plants for the purpose of growing architecture. Climbing plants were targeted and conceptualised as steerable filaments for integration within a pre-arranged diagrid scaffold. Initial investigations focused on strategies of steering plants to shape using decentralised and distributed robot nodes to control stimuli of light and the growth hormone auxin. These studies were then extended to steer plant filaments into structurally performing interlaced configurations, transforming the artificial diagrid scaffold into a bio-hybrid tri-axial Kagome weave. The full system comprised plants, phytosensing devices connected to plants to monitor electrophysiology, scaffolds together with scaffold-producing robotic devices and a distributed, decentralised network of robotic nodes for steering plant growth. In lab conditions, these system elements supported growth towards defined goals, but also embedded the capacity for autonomous self-repair in the face of damage.

The hypothesis for the Co-occupied Boundaries project was that architectural boundaries can be designed to support symbiotic multispecies habitation, in this case between human occupants and bees. 3D printing provided the means for producing porous cellular structures allowing bees to inhabit and extend through their own construction. Component topology and geometry was computationally generated and prepared for 3D printing using a customised print process. The premise of this project has been taken forward into the EU-funded project HIVEOPOLIS which is investigating 3D-print production of healthy hives by incorporating specific strains of mycelium with antiviral properties beneficial to bees.[7]

Flora Robotica consortium,
Flora Robotica,
Centre for Information Technology
and Architecture (CITA),
Royal Danish Academy of Fine Arts (KADK),
Copenhagen,
Denmark,
2016

The project investigated the construction of symbiotic relationships between plants and robots for the purpose of growing architecture.

Asya Ilgun,
Co-occupied Boundaries,
Centre for Information Technology
and Architecture (CITA),
Computation in Architecture Master's programme,
Royal Danish Academy School of Architecture,
Copenhagen,
Denmark,
2016

below: This project investigated the use of 3D printing to construct complex boundaries to support multispecies inhabitation. The introduced bee colony embellished the 3D-print scaffold with their own structures, creating a continually evolving bio-hybrid architecture.

bottom: Stages in the computational design of a 3D-print scaffold. A topology optimisation method was parameterised to ensure that material placement allows bee habitation within the porous structure whilst retaining structural performance capacity.

The Fungal Architectures project has the objective of developing a fully integrated structural and computational living substrate using fungal mycelium for the purpose of growing architecture. The project targets two technological breakthroughs: (1) growing mycelium composites at building scale; and (2) functionalising the living mycelium network to act as a computationally active material. To achieve the first objective, a construction concept has been developed using a stay-in-place reinforcement made using Kagome weave principles. This scaffold provides a breathable frame which is filled with inoculated substrate and grown to create monolithic building elements. To achieve the objective of functionalisation, a number of strategies are being investigated including uptake of nanoparticles by the developing mycelium network, and monitoring resistance changes in the living composite (the composite being the mycelium together with the substrate it is growing in), which is a function of the composite's moisture distribution and conditional on environmental parameters. It is hypothesised that in both cases, informed design of geometries and volumetric parameters of material can be tuned to respectively influence spike-based computation, or parallel computing units using differentiated structural elements to generate and exchange information. The promise of this project is that architectural scale configurations of material and space will produce, or at least influence, their own inherent and tangible computational characteristics.

Claudia Colmo,
Restless Labyrinth,
Centre for Information Technology
and Architecture (CITA),
Computation in Architecture
Master's programme,
Royal Danish Academy
School of Architecture,
Copenhagen, Denmark,
2020

above: Prototype wall section 3D-printed with contaminated soil and inoculated with *Pleurotus ostreatus*. The mycelium fully colonised the wall within 35 days, demonstrating feasibility and biocompatibility.

below: Environmental succession diagram. Bio-remediation provides a foothold for subsequent species to establish and mature, transforming the site into an oasis of biodiversity.

Phil Ayres,
Fungal Architectures
construction concept prototype,
Centre for Information Technology
and Architecture (CITA),
Royal Danish Academy
School of Architecture,
Copenhagen,
Denmark,
2019

Woven stay-in-place scaffolds will act as a combined mould and reinforcement for living mycelium composites to grow and become functionalised as a computationally active substrate.

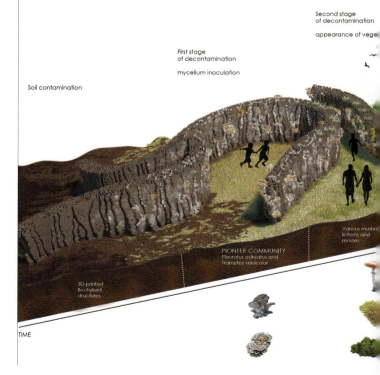

The labyrinth wall geometry was carefully contrived to allow for local modifications. The reticulated geometry provides visual continuity whilst being tectonically composed of discrete elements, allowing for local adjustments to spatial connectivity.

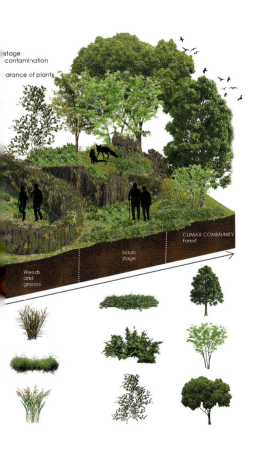

The premise of the Restless Labyrinth project was to explore an in-situ remediation strategy for contaminated sites using fungi as the active decontaminating agent. The feasibility of fungal-based remediation is well studied in the literature, providing mappings between ideal fungal strains for specific contaminants, data on rates of decontamination and protocols of preparation. Studied decontamination rates of approximately one month provided the temporal basis for dynamic reconfiguration of the labyrinth, rendering it restless through the use of 3D-printed architectural boundaries composed from contaminated soil of the site and inoculated with the fungal strains *Pleurotus ostreatus* and *Trametes versicolor*. The feasibility of this proposition, in terms of biocompatibility, material composition for printing and 3D-print parameters, was empirically tested in lab conditions. This testing provided the basis for exploring a tectonic approach that supports spatial reconfiguration through local modification, and the speculation of ecological succession and increasing biodiversity resulting from this bio-hybrid architectural intervention.

Prospects

The study of architectural bio-hybrids promises to uncover deep reservoirs of creative potential that can be brought to bear on some of the most pressing environmental challenges currently facing the built environment. But working with living complexes also offers the opportunity to enrich the palette of orthodox architectural objectives, allowing the extension, or even invention, of architectural programme, function, performance and embodiment of values as demonstrated through the projects described here. The task of architecture to define boundary, frame, filter, stage and to connect has the potential to be radically reimagined through this renewed and invigorated world-building model of biological coupling.[8]

Notes

1. Peter Collins, 'Biological Analogy', *Architectural Review*, 126, 1959, pp 303–6.
2. Freidrich Küffner, *Architectura viv-arboreo-neo-synemphyteutica pomonea, horologica, floralis, hydraulica, sylvestris, fortificatoria, henotica et hypomnematica: oder neu-erfundene Bau-Kunst zu lebendigen Baum-Gebäuden, durch auch neu-erfundene Propff-u. Peltz-Kunst...*, 1716.
3. Arthur Wiechula, *Wachsende Häuser aus lebenden Bäumen entstehend: mit Tafeln*, Verlag Naturbau-Gesellschaft (Berlin), 1926.
4. See www.ferdinandludwig.com/.
5. See https://www.florarobotica.eu/.
6. See www.fungar.eu/.
7. See https://www.hiveopolis.eu/.
8. Flora Robotica was funded from 2015 to 2019 by the European Union's Horizon 2020 research and innovation programme, FET Proactive Action, under grant agreement No 640959. Fungal Architectures is funded from 2019 to 2022 by the European Union's Horizon 2020 research and innovation programme, FET OPEN, under grant agreement No 858132.

Text © 2021 John Wiley & Sons Ltd. Images: pp 96–7 © 2020 Fungal Architectures/CITA, All rights reserved. Modelling framework development by Adrien Rigobello; p 99(t) © 2015–19 Flora Robotica/CITA, All rights reserved. Photography by Anders Ingvartsen; p 99(c&b) © 2016 Asya Ilgun, All rights reserved; p 100(l) © 2019 Fungal Architectures/CITA, All rights reserved. Photography and prototype production by Phil Ayres; pp 100(t), 100–1(b), 101 © 2020 Claudia Colmo, All rights reserved

Kathy Battista

The White Cube in Virtual Reality

Adam Broomberg, Guy De Lancey and Brian O'Doherty,
The White Cube,
2020

The work investigates a virtual white-cube space that is modelled after the museum in Ankara where a Russian ambassador to Turkey was murdered. The work explores O'Doherty's seminal book *Inside the White Cube* (1986) in relation to the virtual gallery space. Originally created to be seen in typical VR fashion with a headset, it was transferred to video during the COVID-19 pandemic for the online exhibition 'Artists and Allies' at the Signs and Symbols gallery in New York.

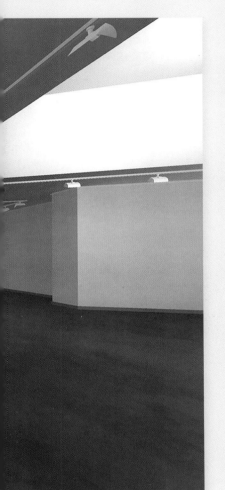

The evolution of augmented, mixed and virtual realities has opened up opportunities for some artists to construct work that in the past would have required expensive sets. Art historian and curator **Kathy Battista** shows us some examples of these new creative models of practice and the speculative worldmodelling they have produced.

As technology has transitioned from the analogue to the digital, art production has seen a concomitant evolution. With the advent of virtual- and augmented reality in recent years, artists have embraced these platforms in their practices, enabling them to build their ideal digital environments without the hassle of constructing sets or models. Where artists such as Dan Graham, James Casebere, Isek Bodys Kingelez and Thomas Demand, to name but a few, used scaled architectural models to create alternative worlds in the last decades of the 20th century, today artists are transitioning to worldbuilding with VR. However, like earlier debates surrounding photography, the politics between artist, apparatus and subject are not fixed, but shifting ideas around issues of agency and exploitation. Works of art made using virtual reality and shown in VR as well as online video and personal device formats create alternative worlds, yet we must question the ethics of such spaces. Is digital colonialism yet another manifestation of the history of art production, the market and the viewer/consumer?

Institutional Support for New Technologies
The use of VR and AR heralded a new era in contemporary art of the 21st century. While VR and AR are not yet ubiquitous technologies, several art institutions began to support new directions in this practice around 2010. New York's New Museum's 'First Look: Artists' VR', originated in 2012 with an exhibition of six artists' VR projects that were experienced by downloading a free app.[1] Aligned with the museum's remit to focus on emerging and cutting artists, younger artists such as Jayson Musson, Rachel Rossin and Jacolby Satterwhite created new pieces for this project. Indeed, while created using VR technology, most visitors will experience the app on their phones or personal devices, and VR headsets are not required. 'Acute art', another app that specialises in VR, has commissioned leading artists including Marina Abramovic, Jeff Koons, Cao Fei and KAWS, among others.[2] In this instance, the creators have relied on blue-chip international artists who are household names, presumably to ensure a wide experience and user set.

In addition to these platforms created specifically for virtual reality, this new medium has also been integrated into larger art exhibitions. Pedro Reyes's major commission for New York City public art agency Creative Time, *DOOMACRACY* (autumn 2016) included a VR room called *Apocalypse Park*, where visitors were transported to a forest with demonic characters wielding chainsaws. A response to the privatisation of parks, Reyes used this technology to produce an immersive experience where a viewer can witness the futuristic horror of parks gone wild.

More infamous is Jordan Wolfson's *Real Violence,* which created heated controversy during its installation at the 2017 Whitney Biennial in New York. Here, viewers (over 18 years of age) used noise-blocking headphones and VR headsets (somewhat unthinkable post-COVID-19) to watch the 2-minute, 24-second piece. *Real Violence* opens on a sunny street in New York where we see two young men of similar demographics: race (white), age (20s) and build (lean). The artist stands on the sidewalk holding a baseball bat, while another young man (an actor whose face is superimposed on an animatronic) kneels on the curb, in a submissive position as if ready for an attack. Wolfson's figure then smashes the other man's skull and proceeds to kick and beat the victim, while blood gushes out of his head. Eventually he resorts to kicking the victim's head, and one sees and hears the bat roll to the curb. The soundtrack of this piece features a man singing a Hebrew prayer that is played during Hanukah celebrations. As Alexandra Schwartz observed in her *New Yorker* review, *Real Violence* felt like 'substituting smooth, crystalline clarity for a video medium that we are more familiar with: the handheld shakiness of a smartphone camera capturing something urgent or horrible as it unfolds.'[3]

Wolfson's recontextualising of violence, from the social and mass media we are drip-fed daily into the supposed sanctity of a white-cube space, is what makes the VR disturbing. What we have since experienced as a society, including the real-time murder of George Floyd by Minneapolis police officers in Spring 2020, sadly seems like a common occurrence.[4] Wolfson's use of virtual reality implicates the viewer into the scene, unable to look away unless a headset is removed, as many visitors felt obliged to do. Indeed, it became a test of one's *chutzpah* to get through the full 2 minutes and 24 seconds of the work. In the examples of Reyes and Wolfson, virtual-reality modelling creates worlds that integrate and overwhelm the viewer, implicating them into the violence depicted within each scenario.

Other artists, for example Sarah Meyohas, have created VR and AR works as independent projects. Meyohas's 'Speculations' series (2015–18) used her own body, a mirror and various props to suggest infinite worlds. These were created in tandem with her *BitchCoin* currency: in order to have value in the currency the artist needed something to be exchanged, thus the ongoing series of photographs. Meyohas also created a VR project, *Cloud of Petals*, for her Red Bull Arts exhibition in New York in 2018.

Meyohas's 'Speculations' series used her own body a mirror and various props to suggest infinite worlds

Sarah Meyohas,
Rope Speculation,
2015

Rope Speculation is part of Meyohas's larger ongoing 'Speculations' series that uses mirrors and constructed props in the artist's studio to suggest infinite cubic forms and space that are metaphors for value and the market. In this individual work, the artist uses a rope mirrored to infinity.

Sarah Meyohas,
Hope Speculation,
2015

In line with her earlier bodies of work, such as *BitchCoin* and *Stock Performance*, the pieces in Meyohas's 'Speculations' series interrogate the rules of aesthetics and market value alongside the artist's role within such spheres. Here, a rolled canvas is shaped into a crude cube and mirrored to infinity.

The White Cube

The White Cube (2020) is a virtual-reality work, its title derived from Irish art critic and artist Brian O'Doherty's 1986 series of three seminal essays, which have become canonical for the study of contemporary art and architecture.[5] Created by O'Doherty with fellow artists Adam Broomberg and Guy De Lancey, this short work (2 minutes, 42 seconds) examines virtual reality, which O'Doherty terms as another 'failed Utopia'.[6] Although created in virtual reality, due to the COVID-19 lockdown most will have first experienced the work on a personal flat-screen while viewing the online exhibition 'Artists and Allies' on the website of the Signs and Symbols gallery based on the Lower East Side of New York.[7]

A pared-down work, *The White Cube* avoids many of the flourishes of other contemporary art ventures into virtual reality that are discussed above. Instead, the 'camera' pans through the bare walls of a gallery, while O'Doherty's voice narrates passages from his text, with additional commentary and even song at one point. O'Doherty's essays, as well as a series of interviews that De Lancey recorded with the 92-year-old artist, are the sources of the words that we hear. The sense of intimacy between the collaborators comes across in the calming sound of the older artist's voice, his laughter, and playful narration.

Broomberg, the originator of the project, describes O'Doherty's essays as one of those truly life-changing narratives that once read can never be undone in one's mind.[8] Alongside, for example, John Berger's *Ways of Seeing* (1972), Edward Saïd's *Orientalism* (1978) or Laura Mulvey's 'Visual Pleasure and Narrative Cinema' (1973),[9] O'Doherty's texts encapsulated the postwar trend of the white-walled, minimal gallery space. They were later published as a book, that has become a de facto bible for contemporary art spaces, as well as having engendered a genre of architecture found in major galleries and museums worldwide, and in addition the ubiquitous art fairs. As we know, the concept of the white cube has surpassed being an art-world phenomenon and has found its way into interior and retail design: Apple, Target, Habitat, even pot purveyors MedMen, have co-opted the look of a so-called 'neutral space'.

The video opens to a rendering of a minimal white gallery. The space is intentionally rendered somewhat rougher than can be achieved with current VR technology, encouraging the viewer to consider the artifice of the medium. The first words one hears after the title credit are: 'Unshadowed, clean, white, artificial, the space is devoted to the technology of aesthetics.' The 'camera' then pans through the white cube and its adjacent spaces that encapsulate O'Doherty's text: windowless, hermetic, mausoleum-like. O'Doherty's seductive Irish lilt lends a sense of humanity to his theories on the white cube: 'The essentially religious nature of the white cube is most forcefully addressed by what it does to the humanity of anyone who enters it. One does not eat, drink, lie down or sleep.'

O'Doherty's words, when heard accompanied by the virtual-reality video, can be considered in parallel to traditional modelling, in which a scaled, miniaturised version of a structure is lifeless, but suggests the life to come in an undetermined future in that space. Small figures often populate a three-dimensional or virtual model to enable a client to understand how users will occupy an intended space. The same is true for galleries who create physical models of art fair exhibitions. However, as every architect, builder or gallerist knows, space can never be controlled and is always subject to change. The failure of International Modernism and its utopian ideals is the perfect example. Architects propose built worlds, but it is the people within them that determine how the space is used. O'Doherty says in the video: 'Space is not just where things happen. Things make space happen.' The collaborators echo this notion within the virtual-reality work. The precise, hygienic space of a gallery is one of the most controlled environments that one might find, second only perhaps to labs or hospitals; yet, the artwork and the people within the space make that world come alive. If a digital rendering of space, by artists or architects, can manipulate our emotions, actions and behaviours, then what is the responsibility of those creating these spaces? Digital colonialisation echoes that of real gallery and museum spaces, most of which rely on plunder looted through colonial campaigns of earlier eras.

The space rendered in *The White Cube* is taken from an earlier iteration titled *Woe from Wit* that was shown, with VR headsets, at the Synthesis Gallery in Berlin in 2019. (*Woe from Wit* is credited to Broomberg & Chanarin, the artist duo that made work together for over two decades, in collaboration with Brian O'Doherty and Guy De Lancey.) This version featured the same premise: a virtual-reality gallery space that adheres to the principles of the white cube as laid out by O'Doherty. *Woe from Wit*, and the space that we see in *The White Cube*, is a mockup of the Ankara museum where the Russian ambassador to Turkey, Andrey Karlov, was assassinated at an official event in

2016.¹⁰ Karlov attended the opening of a show of amateur photographs created by Turkish citizens who had travelled to Russia. In *Woe from Wit* these innocuous photographs are hung in the same position as they were in the Ankara museum. The artists also included shots of Karlov's assassination, as seen from several angles throughout the gallery. Broomberg contacted several of the press that had been present (for a gallery opening, who in turn witnessed a murder) and obtained photographs, anticipating the possibilities that might be afforded by this virtual-reality mockup.

In his narration of what a white cube does to the humanity of anyone who enters it, O'Doherty says: 'One does not get ill, go mad, sing, dance or make love.' The juxtaposition of these words and the violence of the gallery makes overt what is less obvious in *The White Cube*. The latter provokes one to think of all the various insidious violations that happen to all of us in the digital realm. Alexa eavesdrops on our conversations about products that suddenly appear in our Instagram feed. We click 'Accept' without reading Apple's pages of legal documents, unwittingly giving our consent to our entire lives being tracked and recorded and used to sell more things in late-capitalist society. How controlled is this space? And who is in control?

The space that we are taken through in *The White Cube* is devoid of the additional layer of political performativity of *Woe from Wit*. As Broomberg cites, galleries had been the site of violent performances for the past half-century, from Chris Burden's *Shoot* (1971) and Marina Abramovic's *Rhythm O* (1974). Pared down to just the space itself, the later work allows viewers to imagine the potential of the space, for performance, for violence by an artist or otherwise, for any of the actions that O'Doherty says are not acceptable in the white cube. Broomberg says: 'What does *The White Cube*'s space offer as a potential world? It renders a Utopian space where anything is possible, yet it alludes to the violence of technology as an inherent problem of virtual reality: what are the ethics of virtual reality?'¹¹

Galleries had been the site of violent performances for the past half-century, from Chris Burden's *Shoot* (1971) to Marina Abramovic's *Rhythm O* (1974)

Chris Burden,
Shoot,
F Space,
Santa Ana, California,
1971

In 1971, Burden asked Bruce Dunlap, an art student studying on a GI bill and an expert marksman, to shoot him in the upper-left arm as part of this performance. In video documentation Burden is seen flinching after receiving the shot, but he walks away unassisted, unlike so many soldiers in Vietnam that the artist saw shot daily on his television screen. Burden's *Shoot*, typical of his early performances, challenged previously established notions of what was considered art, and questioned the acceptable behaviour of artists and observers within the gallery space.

Decolonising the Colonised

Broomberg's forthcoming project uses augmented reality to comment on the demographics of the Arles Festival of Photography in 2021. Using a specially designed app, viewers can point their phones at a fashion photograph from the 1970s by Guy Bourdin, but on their screens will see an Algerian photographer from the same era, thereby decolonising the colonised space of the art festival. If there are 10,000 photographs in the festival, then each will have an alternative on the app. Credit and payment will be given to all of the artists whose work is featured, a refreshing change from their ongoing exploitation by some elements of the art world. Artistic labour, for 99 per cent of the artists in the world today, is cheap. We can compare this to the underpaid Apple and Amazon workers, sleeping in cars between shifts. Artists will always use technology to subvert the very medium or platform they are creating in. Virtual reality and augmented reality are two examples in a long list of technological developments they have employed. It will be fascinating to see how VR and AR modelling will affect our understanding of art production and viewing practices. ∞

Notes
1. www.newmuseum.org/exhibitions/view/artists-vr#:~:text=%E2%80%9CFirst%20Look%3A%20Artists'%20VR,Rachel%20Rossin%2C%20and%20Jacolby%20Satterwhite.
2. https://app.acuteart.com.
3. Alexandra Schwartz, 'Confronting the "Shocking" Virtual Reality artwork at the Whitney Biennial', *The New Yorker*, 20 March 2017: www.newyorker.com/culture/cultural-comment/confronting-the-shocking-virtual-reality-artwork-at-the-whitney-biennial.
4. Evan Hill *et al*, 'How George Floyd was Killed in Police Custody', *New York Times*, 31 May 2020: www.nytimes.com/2020/05/31/us/george-floyd-investigation.html.
5. Originally published as a series of three essays in *Artforum* in 1980, they can now be found in a book: Brian O'Doherty, *Inside the White Cube: The Ideology of the Gallery Space* [1986], University of California Press (Berkeley, CA), 2000.
6. Brian O'Doherty unpublished interview by Guy De Lancey. Thanks to Adam Broomberg for making this material accessible.
7. www.signsandsymbols.art/exhibitions/artists-allies-iii-b7fq.
8. Broomberg in discussion with the author, 12 August 2020.
9. See John Berger, *Ways of Seeing*, Penguin (London), 1972; Edward Said, *Orientalism*, Pantheon Books (New York), 1978, and Laura Mulvey, 'Visual Pleasure and Narrative Cinema' [1973], in *Visual and Other Pleasures*, Palgrave MacMillan (London), 1989, pp 14–28.
10. 'Russian Ambassador Andrey Karlov Shot Dead in Ankara', *Al Jazeera*, 20 December 2016: www.aljazeera.com/news/2016/12/russian-ambassador-andrey-karlov-shot-ankara-161219162430858.html.
11. Broomberg, email to the author, 4 August 2020.

Text © 2021 John Wiley & Sons Ltd. Images: pp 102–5 © The Artists, Courtesy of Signs and Symbols Gallery, NY; pp 107–8 © Sarah Meyohas; pp 110–11 © Chris Burden / Licensed by the Chris Burden Estate

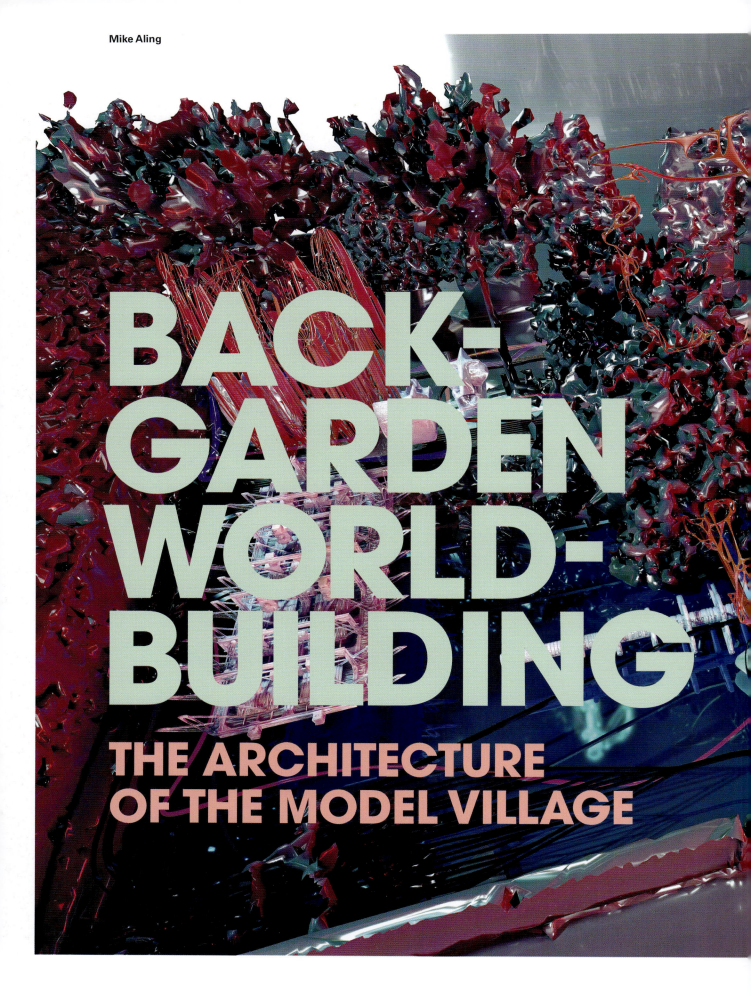

Mike Aling

BACK-GARDEN WORLD-BUILDING

THE ARCHITECTURE OF THE MODEL VILLAGE

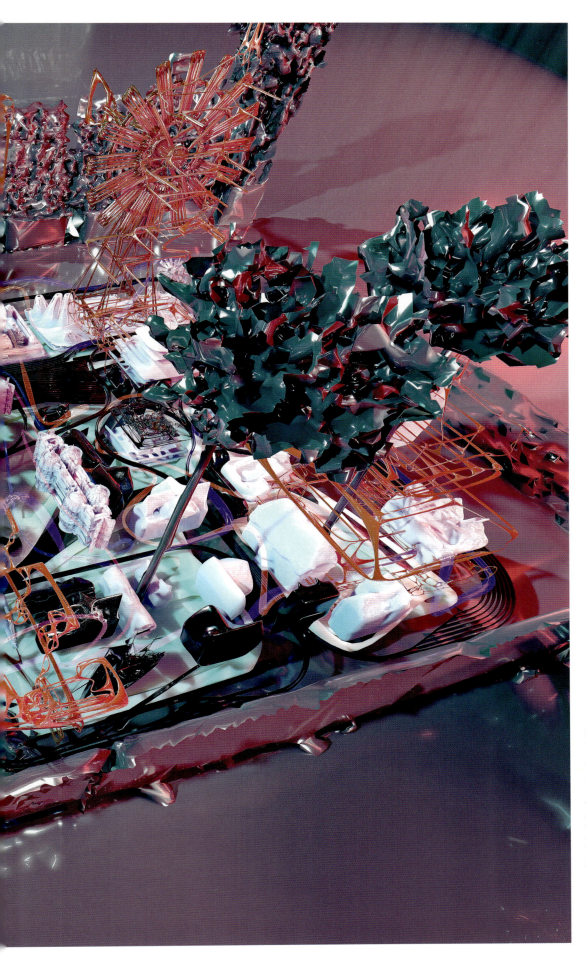

Mike Aling,
Groenwych for
DLR Model Village,
Greenwich,
London,
2020

This village research project in development researches and speculates on how the Greenwich area can be explored as a 'model' space through different uses of the term. It also investigates the long and peculiar histories, eccentricities and tropes of the British model village form. The image shows the model of the Groenwych model village that will be sited in it. This fractal game of nested models has long been played in British model villages.

Guest-Editor Mike Aling describes his infatuation with the model, particularly the long tradition of the British model village and its often-surreal manifestations. He focuses on some of his favourites and introduces us to his own contemporary model village – Groenwych.

Mike Aling,
Extant model villages in the UK,
2020

Plans of nine surviving British model villages, outlining their relative areas, circulation patterns, model railway tracks and water features.

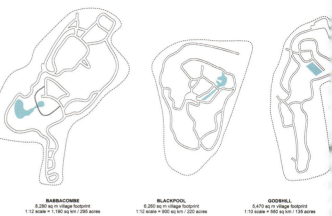
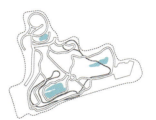
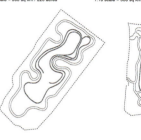
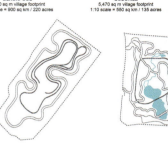

British model villages had their golden age as popular tourist destinations from the 1930s to the 1960s, and around 20 are still in existence today. The model village is a curious British invention and although each is unique, they share many traits: they are of similar sizes and scales (often 1:12 or thereabouts); have a range of similar building types, forms and programmes; and are invariably nostalgia machines that lament for times past. The term 'model village' here of course refers to a miniaturised settlement, rather than the other distinct use of the term describing idealised full-scale towns such as New Lanark in Central Scotland, Bournville in the English Midlands or Portmeirion in North Wales.

Model villages are located in wonderfully banal types of outdoor sites including garden centres, the grounds of larger visitor attractions, and beach promenades, but gardens are by far the most common site. This might be the garden of a pub looking to attract additional trade, such as the model village at Bourton-on-the-Water (1937) in the English Cotswolds, or the private garden of the model-village maker that opens itself to the public (Bekonscot (1929–) in Beaconsfield just to the west of London), or the garden of a disinterested landlord (the now-defunct model village at Ramsgate in Kent, England (1953–2003)).[1] While the practice of worldbuilding is mostly an indoor endeavour, where worlds are developed in the confines of offices, studies or bedrooms – perhaps now more than ever, during the COVID-19 pandemic at the time of writing – model villages have long imagined worlds in the context of the back-garden site.

114

Model Villaging as Worldmodelling

Model villages can be determined as a separate typology to their close relation the *miniature city*.[2] A straightforward differentiation is that model villages are sited outdoors, whereas miniature cities are indoor enterprises with carefully controlled atmospheres. Also, often miniature cities are raised to near eye level, either on tables or through accentuated landscaping, whereas model villages are firmly sited on the ground.

The model village is sited in the not-quite real, or more often the never-was real in their particular form of toy-like simulacra. Sometimes, they are yet-to-be real, and share this condition with architectural models. Unlike the architectural model however, model villages are largely functionless: they are a pursuit of delight. They are also worldbuilding exercises, and like architectural models, their success lies in their immersive potential, along with the coherence of their spatial and cultural logic under their own terms of engagement. A cognitive displacement is made by the visitor, not only through scale shifting, but into the strangely familiar world beneath the knees, a supposedly brighter place if the visitor is willing to agree that a more anodyne and apolitical village is an improvement on actuality. Through this the model village exudes a very British form of Magical Realist worldbuilding. This of course is not without its problems. Although shrouded in light-heartedness and a knowingly retrograde mindset, model villages emanate overly prescriptive idealisations of rural communities, steadfastly affirming quintessential English village-ness from the inter-war years of their early golden age. Although seen as a safely distant anachronism today, it must have seemed a disturbingly conservative ideology in the post-war period (its late golden age), given the tremendous social, political and civil changes of that time.

The garden is the worldspace of model villages. There is a curious sense of a diminutive picturesque at play in many model villages, one that actively necessitates the visitor to walk through its miniature urbanism as a core circulatory method. The giganticism felt has inspired numerous apt references to Lilliput in the history of model villages,[3] as we indeed become the 'Quinbus Flestrin' ('Man Mountain') of a 12:1 Lemuel Gulliver.[4] The use of the body as a measuring device offers an alluring strangeness, as Susan Stewart states in her book *On Longing* (1984): 'The body is our mode of perceiving scale and, as the body of the other, becomes our antithetical mode of stating conventions of symmetry and balance on the one hand, and the grotesque and the disproportionate on the other.'[5] Our bodies are demonstrably the wrong scale, yet the willing suspension of disbelief allows us to enter its world. This close proximity of the visitor/occupant to the models allows for the admiration of the craft on display, and for the 'finding' of moments in the miniature community to project oneself into.

Polperro Model Village,
Cornwall,
England,
2020

Polperro Model Village (1948–) is scaled to 1:12, and like the model village at Bourton-on-the-Water, it is a miniature facsimile of the town it is sited in. Polperro tackles the original subject matter of the model village, initially explored in Charles Paget Wade's Wolf's Cove (a development of the first model village): the Cornish fishing village. The building models are mostly made from cement and are repaired annually out of peak season.

These devices are ways that model villages invite us into their communities, through the lives of their occupants and their details, their rituals and events, forever frozen. At times we are offered glimpses into their closest secrets: interior spaces, highly cherished as model villages are almost entirely exterior experiences.

This close proximity affords us a way to understand how the model village can be seen as analogous to the contemporary digital architectural model: through offering the viewer an immersive and multi-scalar experience. Unlike pre-digital physical architectural models which were largely understood as externalised objects that encouraged a distance between the viewer and the architectural subject, this proximity allowed the model village to become a proto-digital model space, unlike the miniature city, kept at a careful distance from the viewer.

Origins of Unparalleled Fancy

The first model village was the product of an architect's imagination, begun in 1907 by architect Charles Paget Wade whilst living in and working on the Hampstead Garden Suburb in northwest London as a junior architect for Raymond Unwin and Barry Parker.[6] Named 'Fladbury', it was designed and painstakingly constructed by Wade in the rear garden of his lodgings at Temple Fortune Hill in Hampstead, for Betty, the daughter of his landlords.[7] Fittingly, Hampstead Garden Suburb is somewhat of a model village itself, in both senses of the term. Concepts around the Garden City Movement championed by Unwin at the time, along with Wade's frustrations in his daily work, fed directly into his evening passion for worldbuilding through modelling.[8] Fladbury predates the Bekonscot model village by over 20 years. It is often misinterpreted as the earliest in the UK, whereas it is the oldest *surviving* model village.

Amongst other obsessions, Wade was a historical costume enthusiast, and was often labelled an eccentric; however, as his biographer Jonathan Howard states, 'the term "eccentric"

Fladbury Model Village in the garden of 9 Temple Fortune Hill,
Hampstead Garden Suburb,
London,
c 1908

Charles Paget Wade constructed the first known British model village, titled 'Fladbury', whilst lodging and working at the Hampstead Garden Suburb. The village was built as a present for Betty, the daughter of Wade's landlords at the time. Fladbury was later relocated and developed into the more ambitious 'Wolf's Cove' at Snowshill Manor in Gloucestershire, England.

115

can serve to belittle or dismiss Charles' determination and achievements, and foreclose any effort to delve more deeply into his values and motivations'.9 The sentiment in this quote can perhaps be seen as true of any model village, let alone its creator: their messages are often disregarded as folly. And it is true that Fladbury was essentially a toy. One must question who the toy is for, whether it is the children that benefit most from the visit to the village, or the parents. The inspiration for Fladbury, like many other model villages, was in the garden model railway sets gaining popularity for wealthy (mostly) men at the turn of the 20th century: these models offered the chance to construct a rural idyll and revive one's childhood in the comfort of the garden. They continue today with colossal dimensions, as outlined by Mark JP Wolf earlier in this issue (see pp 22–31). Initially, rural settlements were built sparsely along the track to conjure lengthy journeys, equating to a greater sense of remoteness from the everyday. Although buildings played only a supporting role at the start, these architectures eventually evolved into the splinter typology of our focus here.

Betty grew up, Wade moved on, and Fladbury was relocated, renamed and vastly expanded into 'Wolf's Cove' in the grounds of Snowshill Manor in Gloucestershire. Wade purchased the Tudor manor house on his return from the First World War, and it became his life's work as he restored and transformed the manor into a curated museum for his extensive collection of objects and artefacts. Wade gifted Snowshill to the National Trust towards the end of his life in the early 1950s and it can still be visited today, with reconstructed elements of Wolf's Cove making seasonal appearances in the grounds, and original elements on display inside. In 1931 John Betjeman visited Wolf's Cove and was so enamoured that he penned a half-joke, half-polemic homage to the model village as an article for the *Architectural Review*, of which he was editor at the time. Betjeman wrote of Wolf's Cove as though it were a full-scale village, describing its everyday rural culture and speculating on the origins of its vernacular architectures.

Charles Paget Wade at Snowshill Manor,
Gloucestershire,
England,
early 1920s

Charles Paget Wade, here photographed outside his home-museum Snowshill Manor in the early 1920s, dressed in Cromwellian-era military attire.

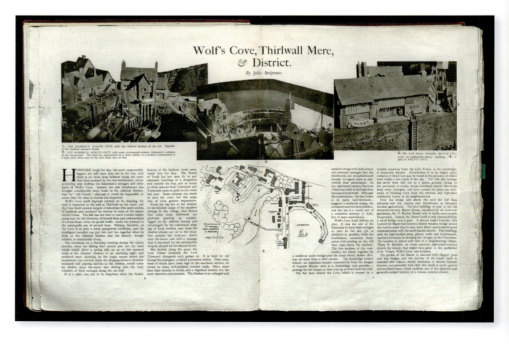

John Betjeman,
'Wolf's Cove, Thirlwall Mere,
and District',
in *Architectural Review*,
January 1932

John Betjeman visited Charles Paget Wade at Snowshill Manor in 1931 and was so enamoured with the Wolf's Cove model village that he wrote an article in the *Architectural Review* – of which he was then editor – imagining the model as a full-scale, inhabited village.

The Contemporary Model Village

Model villages were a 20th-century invention, but their legacy remains etched on the cultural psyche. Most recently, artist James Cauty has been subverting the model village and its tropes in his *ADP (Aftermath Dislocation Principle)* (2013) trilogy of projects, a village distinctly in the dystopian register. Cauty imagines an urban stretch in Bedfordshire peppered with vigilante law, militia justice and breakouts of rioting, all modelled exquisitely at 1:87. Cauty's newest and largest-scale model village, titled *ESTATE* (2020), is currently touring the UK in a shipping container, and takes the chaos and craft of the earlier work to new scales (1:24). Somewhat ironically for a model depicting scenes of destruction, ruin and disaster, it is an exquisite and lovingly made hyperrealist construction.

In my own research I am developing a model village, titled 'Groenwych for DLR'. A work in progress, it is sited on a roof garden in Greenwich, London, and – like all model villages – is both a register of its site context, and a chimera that amalgamates elements found in the canon of model villages. To enter Groenwych, and to offer a window into the 'Digital Doppelvillage' – Groenwych's digital twin where the village's interiors exist – a portable filming apparatus is used in the form of a walking cane that employs miniature cameras to produce a three-dimensional image in a virtual-reality headset. The cameras adjust the user's pupillary distance and

Somewhat ironically for a model depicting scenes of destruction, ruin and disaster, it is an exquisite and lovingly made hyperrealist construction

James Cauty,
ESTATE,
2020

The *ESTATE* project is a dystopian model village artwork currently touring the UK in a shipping container. Made up of four tower blocks, each with their own devastating programme (a live-work-die tower, a children's prison, a care home for the dying, and a centre for neo-pagans), the project subverts the twee image of the model village and makes the subject matter even more monstrous in the process.

Groenwych is a research project that both theorises and offers an immersive engagement with how model villages function as models beyond the act of miniaturisation

Mike Aling,
Variations on a Walking Cane,
Groenwych for DLR Model Village,
Greenwich,
London,
2020

The walking cane is a gateway into Groenwych model village and its virtual-reality interiors. An inverted periscope, it shrinks the user's perception – both height and pupillary distance – to the required scale (1:24 / 1:18 / 1:12 / 1:10 / 1:9), by way of stereoscopic miniature camera equipment at its base that connects to a virtual-reality headset. Its design draws on Charles Paget Wade's cane collection and its colourway is based on the 'Wade Blue' he invented.

Model of the model village at Bourton-on-the-Water, in the model village at Bourton-on-the-Water, Gloucestershire, England,
2019

The '1:9th Wonder of the World' and the only model village with Grade II listed status in the UK, the model village at Bourton-on-the-Water was constructed entirely from local Cotswold stone by craftsmen in the mid-1930s. It is sited near the centre of the 1:1 picturesque village, and in turn this image shows the 1:81 model of the Bourton model village, sited inside the model village. If one looks closely, an even smaller 1:729 model of the model village can be found within.

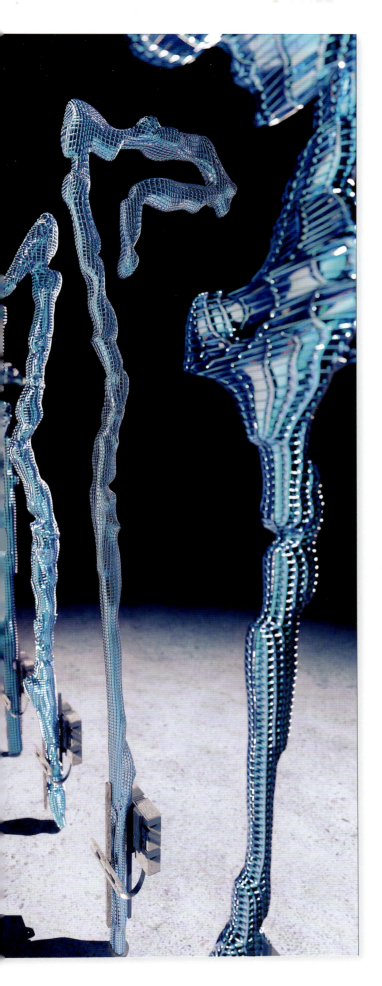

overall height, in order for him or her to experience the space of the model stereoscopically and at a reduced scale. Findings suggest that as the body shrinks and the eyes get closer together, one's perception of depth scales proportionally: a scaled architecture is always scaled to its user. The walking cane is a digital carving made from simulated Banyan vine found on the Island of St Kitts in the Caribbean, home to the late Charles Paget Wade and inspired by his cane collection. Its colourway is matched to 'Wade Blue' invented by Wade.

In keeping with tradition, Groenwych is organised around the parametrics of model railway 'trackplans' (in this instance the southeast London rail network), developed to the dimensions prescribed by railway gauges, including the niche and philosophically on-trend OOO gauge. Trackplanning is undertaken on specific hobbyist software, a model railway enthusiast's equivalent to building information modelling (BIM), with necessary parts and gauges catered for. Groenwych also has its own model village nested within itself, an endless series of fractal villages in reference to those found in model villages such as 'Godshill' (1952–) on the Isle of Wight, England, and Bourton-on-the Water.

Groenwych is built up of models that reference its local real-world architectures. These architectures are curated from throughout the history of Greenwich and studied due to their model-like qualities and intentions, where clear-cut determinations of model and subject become blurred such as in 'The Fame', a full-scale model training (non-)boat moored at Greenwich in the 19th century. Examples of facsimile spaces are also explored, such as the polluted Victorian Greenwich Beach, historically without any discernible beach-like qualities. Local programmes from the area are also surveyed as models-as-exemplars or paragons, including the ideal architectures of Inigo Jones's Queen's House (1635) and Christopher Wren and Nicholas Hawksmoor's Royal Hospital (1742). Spaces that embraced experiences of alterity are also posited – models-as-alterants – including the Palace of Varieties and the outrageous parties at Blackheath Pagoda.

Groenwych is a research project that both theorises and offers an immersive engagement with how model villages function as models beyond the act of miniaturisation. They are spaces of worldbuilding practice that have long afforded us an immersive, multisensory, multi-scalar and politically complex set of experiences, at times to monstrous proportions. This is something they share with today's digital and virtual models. For such a retrograde and often backward-looking enterprise, model villages were strangely ahead of their time. ∆

Notes

1. Brian Salter, *Model Towns and Villages: In Britain … In Public … In All Weathers*, In House Publications (East Grinstead), 2014, p 22.
2. Tim Dunn, *Model Villages*, Amberley Publishing (Stroud), 2017, p 6.
3. *Ibid*, p 16.
4. Jonathan Swift, *Gulliver's Travels* [1726], The Folio Society (London), 1965, p 38.
5. Susan Stewart, *On Longing: Narratives of the Miniature, the Gigantic, the Souvenir, the Collection* [1984], Duke University Press (Durham, NC), 2007, p xii.
6. Paul Capewell, *Charles Paget Wade Before Snowshill: His Early Life and Work at Hampstead Garden Suburb*, On the Road Again (London), 2nd edn, 2019, p 29.
7. Jonathan Howard, *A Thousand Fancies: The Collection of Charles Wade of Snowshill Manor (The National Trust)*, The History Press (Stroud), 2016, p 44.
8. Capewell, *op cit*, p 47.
9. Howard, *op cit*, p 8.

Text © 2021 John Wiley & Sons Ltd. Images: pp 112–14, 118–19 © Mike Aling; p 115 © National Trust / Claire Reeves & team; p 116(t) © National Trust / Sarah Allen; p 116(b) Courtesy of Mike Aling; p 117 © Image courtesy of James Cauty

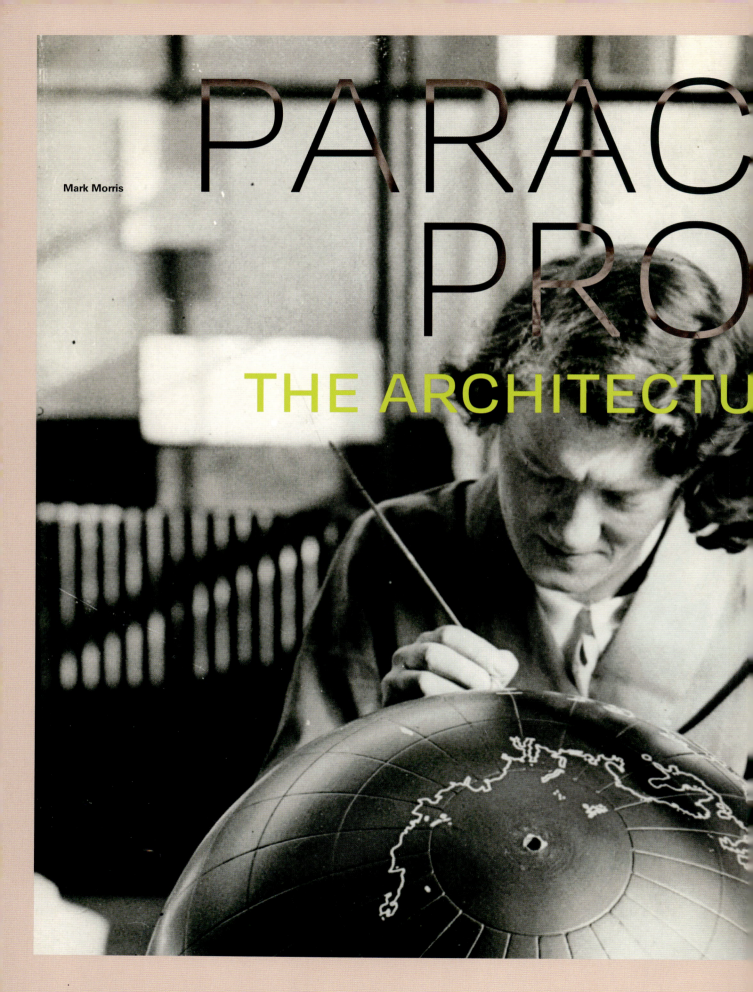

Mark Morris

PARAC
PRO
THE ARCHITECTU

COSMIC
PROJECT
ARCHITECTURAL LONG GAME

A woman painting a globe, c 1930

The making of world models, in this case a literal globe, is not so different from the making of models in general. Douglas Adams created the character Slartibartfast, designer of whole planets. The joy of his role in *The Hitchhiker's Guide to the Galaxy* is that, despite the impossible scale and complexity that would come with such a job, Slartibartfast talks as any architect contending with the design of a single building; there is no discernible difference, and questions of style and authorship remain intact.

The notion of the 'paracosm' is a useful one for architects to understand. Some of the best writers and designers often work within personal, modelled imaginary worlds, sometimes constructed over decades, enabling them to give birth to their narrative scenarios and spaces. Here, Guest-Editor **Mark Morris** looks at various literary and architectural paracosmic precedents.

Look at me: I design coastlines, I got an award for Norway. Where's the sense in that? … I've been doing fjords all my life. For a fleeting moment they become fashionable and I get a major award. In this replacement Earth we're building they've given me Africa to do and of course I'm doing it with all fjords again because I happen to like them, and I'm old-fashioned enough to think that they give a lovely baroque feel to a continent. And they tell me it's not equatorial enough.

— Slartibartfast in Douglas Adams's *The Hitchhiker's Guide to the Galaxy*[1]

The term 'paracosm', a neologism from the 1970s, is used to describe a detailed and consistently structured imaginary world with its own cultures, history, religions, geography, weather and even languages. By definition, a paracosm originates in childhood fantasy and may persist into adulthood as a personal creative laboratory. It can be fantastic or based largely on reality. Scholars have linked the notion of paracosm to certain literary works and authors: the Brontë siblings' shared worlds of Angria and Gondal, Hartley Coleridge's Ejuxria, CS Lewis's Narnia, Austin Tappan Wright's Islandia, Muhammad Abd-al-Rahman Barker's Tékumel and Joanne Greenberg's Kingdom of Yr. Psychologists tend to highlight paracosmic tendencies as coping mechanisms used by children who have experienced a profound loss in an effort to gain control through a surrogate reality.

The loss of their mother to cancer in 1821 was the likely shared trauma that bound Charlotte, Branwell, Emily and Anne Brontë up in extraordinarily detailed imaginary diversions triggered by a gift of toy soldiers to Branwell. These 'young men' were claimed by his siblings as characters in tales of conquest and drama that involved maps, minutely written diaries and gazettes, and instantaneous shared spoken stories taken up by one sibling and passed to the other. The first stories grew out of model play, setting up the soldiers in so many scenes aided by another gift, a toy village and other props. The link to the toys faded away as the narratives became more embroidered. Charlotte and Branwell had initiated an imaginary venue called Glass Town that Emily and Anne were permitted to join; this evolved and split into whole 'worlds', Angria and Gondal, shared by pairs of siblings. These worlds served as incubators for later mature writing from the Brontës. There would not be a *Wuthering Heights* nor a *Jane Eyre* nor a *Tenant of Wildfell Hall* had there not been an Angria or Gondal.[2]

As a directed form of imagination, a paracosm nurtures the intellect, yielding discoveries with real-world application. A 2006 Michigan State University study found a disproportionately high number, relative to the general population, of MacArthur Fellows (so-called 'genius grantees') admit to maintaining paracosms. Michelle Root-Bernstein asserts in the *Creativity Research Journal* that paracosms 'supplement objective measures of intellectual giftedness …

Branwell Brontë, Pillar Portrait, c 1835

The Brontë siblings – Anne, Emily and Charlotte – with the figure of Branwell turpentined away and painted over as a pillar. All four would share paracosms. These would be described in so many miniature books and newspapers, maps and drawings.

as well as subjective measures of superior technical talent'.³ The theme of paracosm may be a useful guide in examining other worlds, civilisations of science fiction, and the history of architecture. Paracosmic thinking could be considered a valid methodology for the intellect, a basis for worldbuilding as one form of output. Rather than view paracosms as merely coping mechanisms, a thesis may be put forward claiming that paracosmic thinking may be a defining attribute of many architects. Architectural vision can be sourced from a persistently tended interior world.

Contrasts to Dreamlands

Time is a key element with paracosms. Time might be sped up or slowed down to work out the impact of one decision on the next following cause-and-effect logic. Time can also be times, multiple worlds in dialogue. Augustus Welby Pugin's illustrations for *Contrasts: Or, A Parallel Between the Noble Edifices of the Middle Ages and Corresponding Buildings of the Present Day* (1836; revised edition 1841) present his idealised perception of 15th-century largely ecclesiastical architecture contrasted with his cynical perception of the architecture of his day.⁴ Both the past and present conceived by Pugin are his worlds, each bound to its own logic; Catholicism versus the Industrial Revolution. At the same time that *Contrasts* was being written and illustrated, Pugin was helping Charles Barry with the interior design for a design competition entry for the new Houses of Parliament at Westminster. The competition was won the same year that *Contrasts* was first published. Pugin's illustrations for his book aided those he did for Barry and vice versa. Pugin was never going to see Britain return to medieval architecture in the way he wanted, but by directing his mania for it towards a speculative project that started with *Contrasts* and matured into his *True Principles of Pointed or Christian Architecture* (1841),⁵ he secured his reputation and saw fragments of his paracosm built.

Another example of an architectural paracosm is Frank Lloyd Wright's Broadacre City. Wright worked on this utopian democratic world from the 1920s to his death in 1959. As with the Brontës, the origins of Broadacre were prosaic, but as the years passed the concept broadened and became more speculative. In his 1932 book *The Disappearing City*,⁶ Broadacre City is still coming together as an architectural and anti-urban planning idea, but this would soon thereafter become a series of drawings and, famously, a 12-by-12-foot (3.7-metre-square) scale model fashioned by his students at Taliesin in Wisconsin and Arizona. This would be internationally exhibited and more 'collateral models'

Augustus Welby Pugin, Frontispiece engraving for *Contrasts: Or, A Parallel Between the Noble Edifices of the Middle Ages and Corresponding Buildings of the Present Day*, 1841

Showing architecture – 19th-century eclecticism versus the Gothic – in the balance, the scales indicate that Gothic architecture weighs more in truth. *Contrasts* offered up Pugin's imaginary medieval world as the antidote to so many sociopolitical, economic and cultural problems of his own day. Pugin did not just feature architecture, but illustrated narratives and commentary alongside his cathedrals and monasteries.

123

Lebbeus Woods, Region M, 1984

In his 1993 manifesto, Woods declared, 'I am an architect, a constructor of worlds'. He had taken this stance from the start of his career as evidenced by projects like Region M. Images like *London Article* couple Woods's architectural draughtsmanship with that of an illustrator. They are particularly arresting owing to the combination of these graphic sensibilities.

For Wright or Woods, the paracosm collects projects and gives them 'gestational life' in the imagination. The model-ness or suggestion of working to a small scale in their paracosms offers a ready form of critical distance; the small seeming far away and more readily graspable would follow. Wright would return to Broadacre City on and off for the rest of his life, updating and perfecting it. This tinkering was carried out conceptually, in drawing and, notably, by revising and making new models. Broadacre City would become a repository for speculative projects, including the mile-high (1.6-kilometre tall) 'Illinois' tower (1957) and his fanciful take on a helicopter (1958), as well as an archive of designs he had realised. Everywhere and nowhere in his practice, Broadacre City spawned designs that were built and conserved many more that were not.

One could find similar elements in projects like Bruno Taut's Alpine Architecture (1917), Le Corbusier's Ville Contemporaine (1922) or Archigram's Walking City (1964). All three began more as polemical statements, manifesto works, but each held a longer and larger intellectual project for the architects involved. These were glimpses into parallel worlds. As a genre of science fiction looks to 'alternative histories', these architectural statements were alternative futures. Lebbeus Woods made a whole career of such paracosms. Projects spanning the 1980s like Centricity, A-City, 4 Cities and Beyond, and Region M were consistently imagined places where Woods could go in order to further different sorts of architectural research. As a co-founder of the Research Institute for Experimental Architecture in New York City in 1988, Woods championed a rare breed of intellectual engagement with architectural questions including societal, political and environmental issues, and supported the work of others – students and colleagues – in this direction. Woods's discourse was exceptional in that it sought out, alongside texts, so many remarkable visualisations of these questions as maps and drawings which served as definitive statements. Notably, the Region M (1984) series of drawings is different from the rest, a paracosm populated by illustrated figures. These peopled images are arresting, almost jarring. They are not minor figures supplied to note scale. They occupy the foreground and tell a story. There is something cinematic about Region M, something reminiscent of graphic novels. There is also the recurring image of globes throughout depictions of Region M, insignia for Woods's worlds within worlds.

Certain contemporary works of sculpture that incorporate architectural models deal with paracosms. Bodys Isek Kingelez's Extreme Models are a series of fantastic and utopian cities. Crafted from cardboard, paper, tinfoil, bottle caps and plastic, they draw inspiration from specific places like Kinshasa, but offer a future vision for African cities in general. Kingelez stated his agenda: 'I make this most deeply imaginary, meticulous and well considered work with the aim of having more influence over life. As a black artist I must set a good example by receiving the light which pure art, this vital human instrument, kindles for the sake of all.'[7]

He made some 500 models in his career; they started out as individual buildings, but he later began to assemble these as sprawling cities with parks, monuments and sports venues. Some of these became major works like Ville Fantôme (1995) and La Ville du Futur (2000), the former exhibited at the Centre Pompidou in the 'Dreamlands' exhibition (2010).

Remembering When
In some respects a paracosm is a *method of loci*, an expanded form of a mind/memory palace put to uses beyond just conserving memories. Frances Yates in *The Art of Memory* (1966) traces the method of recollection from the ancient Greek poet Simonides of Ceos to Cicero to Dante.[8] Venues for these imaginary places built to keep or trigger memories range from a single room where every object represents a memory to whole cities where streets and squares organise clusters of memories each held in a building. Even 'larger', memory systems based on 'Spheres of the Universe' or cosmic orders, adopted by Dante as layers of hell, which uses worlds encapsulated by worlds for layers of memories. Yates views these methods as sitting between rhetorical and philosophical practices and rules for places versus rules for images. Architecture brings all this together, a form of syncretic spatial mnemonics, as a matter of routine. Any architectural paracosm is a method of loci without even trying. For Wright or Woods, the paracosm collects projects and gives them 'gestational life' in the imagination. The model-ness or suggestion of working to a small scale in their paracosms offers a ready form of critical distance; the small seeming far away and more readily graspable. Thus the works committed to memory can take on a life of their own and evolve, interact, and promulgate more designs.

The Picture Gallery
at Sir John Soane's House and Museum,
London,
1830

The gallery – full of Hogarths, Canalettos and Piranesis – is a magical room where walls cleave to reveal more walls, more paintings. A space for paracosmic and world-model thought, the lithograph shows a model set on a table and another at the foot of a sculpture.

Sir John Soane's house museum in London has many attributes of a memory palace, except that it defies the premise by being outwardly expressed and made real. The Soane Museum is all about memory and collapsing an architect's lifetime's work into a larger space of recollected travels and travails. And models play a pivotal role in this world overlooking Lincoln's Inn Fields. Models of built and unbuilt projects, models as souvenirs, models for education; and all of these folded into displayed casts from architecture and sculpture, collections of paintings (most featuring architecture) and a house that literally and figuratively unfolds like a waking dream. In the words of John Elsner:

> As a collection, the models come to figure their collector's desire. Their yearning (and his yearning for them) is to be a complete series (even a series that may repeat some of its members). And so they commemorate the famous objects in which Soane traces his architectural ancestry not only item by item, but also as a totality, a paradigmatic classical heritage.[9]

The Soane Museum qualifies as a built paracosm perhaps even more than it does as an inverted memory palace. It was painstakingly built up in stages over decades (1792–1837) and, like Broadacre City, grew in size and complexity over time, functioning as a design incubator and simultaneously as an exemplar of the architect's total vision. And, as it happened, there was a sense of profound loss that moved it from being a house to a museum owing to both of his sons' behaviour and the impact of their actions on his wife Eliza's health.

Cornelis Cort (after Maarten van Heemskerck), 'Triumph of the World' from *The Cycle of the Vicissitudes of Human Affairs*, 1561

Ingredients for any world: sun and moon; earth, air, fire, water; four winds; and time driving all this. Cort was apprenticed as an engraver in the Netherlands and spent much of his career in Venice and Rome. He was particularly attached to Titian, living in his house and making engravings based on his work.

JJ Grandville, 'Planetary Bridges' from *Another World*, 1844

Surely Grandville was a paracosmicist. Charles Baudelaire, writing in *Présent* in 1857, felt unnerved by such images: 'When I enter into Grandville's work, I feel a certain discomfort, like in an apartment where disorder is systematically organised, where bizarre cornices rest on the floor, where paintings seem distorted by an optic lens, where objects are deformed by being shoved together at odd angles, where furniture has its feet in the air, and where drawers push in instead of pulling out.'

Paracosmic projects – literary or architectural – share several qualities, the most striking of which is how productive and ongoing a nurtured inner world can be for the purposes of creative outputs. If a paracosm is the product of what one might call 'reposed thinking' – something between daydreaming and task-based focus – it would seem a particularly productive vehicle capable of spawning multiple or richly detailed designs that effortlessly interrelate owing to its established backstory. Worldmodelling requires the same depth and exhibits similar attributes. A problem arises when paracosms are assumed to be rare or the domain of geniuses, as if such thinking ability is hardwired for only a few or the result of some horrible personal loss. Would it not be reasonable to assert that such ability could also be cultivated or trained, particularly in an architectural direction? The education of an architect centres on studio assignments that can span a full academic year. A studio brief is a call to engage in a long-term imaginative exercise that synthesises research and historical references and projects innovative proposals, and this training is repeated across four to five years. Architects are effectively trained to be paracosmicists. Claiming architectural vision, having the ability to readily draw a design response by drawing from some *interior portfolio*, is a form of design method. Developing such a method project by project is one option, applying a method across many projects another. Worldmodelling is an architectural paracosmic method. The success of any given world model is had in its own pervasiveness, internal logic and innovative stance cohesively communicated. It's not that a world model needs to be believable, but it must, as it were, believe in itself. 1

Notes
1. Douglas Adams, *More Than Complete Hitchhiker's Guide*, Wing Books (New York), 1989, p 128.
2. Alison Gopnik, 'Imaginary Worlds of Childhood', *The Wall Street Journal*, 18 September 2018.
3. Michelle Root-Bernstein and Robert Root-Bernstein, 'Imaginary Worldplay as an Indicator of Creative Giftedness', *Creativity Research Journal*, 18 (4), Taylor and Francis (London), 2006, pp 407–16.
4. A Welby Pugin, *Contrasts: Or, A Parallel Between the Architecture of the 15th and 19th Centuries*, self-published, 1836; republished with revisions as *Contrasts: Or, A Parallel Between the Noble Edifices of the Middle Ages, and Corresponding Buildings of the Present Day, Shewing Present Decay of Taste*, Charles Dolman (London), 1841.
5. A Welby Pugin, *The True Principles of Pointed or Christian Architecture*, John Weale (London), 1841.
6. Frank Lloyd Wright, *The Disappearing City*, William Farquhar Payson (New York), 1932.
7. Bodys Isek Kingelez, 'The Art of the Model – An Erudite Art', manifesto manuscript, 2002, Estate of Bodys Isek Kingelez.
8. Frances A Yates, *The Art of Memory*, Routledge & Kegan Paul (London), 1966.
9. John Elsner, 'A Collector's Model of Desire: The House and Museum of Sir John Soane', in John Elsner and Roger Cardinal (eds), *The Cultures of Collecting*, Reaktion Books (London), 1994, p 173.

JJ Grandville, 'The Good God' from *The Complete Works of Béranger*, 1836

From a theatrical family, Jean Ignace Isidore Gérard Grandville moved from Nancy to Paris to start his career as a caricaturist – his work often linked to political satire – and book illustrator in his early twenties. Popular in his own time, his work would greatly influence the Surrealists.

Text © 2021 John Wiley & Sons Ltd. Images: pp 120–1 © Mary Evans Picture Library; p 122 © Mary Evans / SZ Photo / Scherl; p 123 courtesy AA Archives; p 124 © Estate of Lebbeus Woods. Image courtesy Friedman Benda; p 125 Wellcome Trust. Creative Commons Atrribution 4.0 International (CC BY 4.0); pp 126(b), 127(b) CC0 1.0 Universal (CC 1.0)

FROM ANOTHER PERSPECTIVE

A Word from
AD Editor Neil Spiller

A Surrealist Rococo Master
Kris Kuksi

Kris Kuksi,
Prosperity,
2020

Kuksi's sculptures, often wall mounted, are highly complex constellations of found and doctored objects. They tell wry, allegorical stories, engage with mythology and leave much to the viewer's reading of them.

A clamping bench on engine parts on patient human feet. At the tip of it all, an old man's too big bearded face looks down at him with obscure curiosity. In his beard a steam train the size of a cudgel, its chimney venting smoke into the bristles.
— China Miéville, 2016[1]

Let us not mince words: the marvellous is always beautiful, anything marvellous is beautiful, in fact only the marvellous is beautiful. — André Breton, 1924[2]

Wayward finials rise into the air. Bewhiskered dragoons, lancers, cavalrymen and hussars from the 19th century jostle for space with voluptuous reclining nudes, mythic muses and surreal chimeras. The main figures in these tableaux are accompanied by a swarm of malevolent or horny putti and steampunk cherubs perched on the larger characters like demented familiars. Guns, swords and hatchets are everywhere. Angels' wings sprout from the most unlikely places, screaming mouths and eye sockets are used as another opportunity of further ghoulish inhabitations. Time in the work is tangled like Mad Max channelling Hieronymus Bosch, all against backdrops that look like they are bashed out on an anvil by a maverick, Victorian blacksmith or designed by a demented architect. Welcome to the wonderful world of American artist Kris Kuksi. A world that has the delicacy and complexity of coral, that is at once disturbing yet sexy and gorgeously beautiful.

The Line of Beauty
Models abound, from the toys we play with as children, to the hobbies some of us have as adults, to the virtual facsimiles we use to illustrate our ideas. Kuksi is a collector of models, plastic soldiers, toy weapons, bears, pigs, gladiators and suchlike, but also ornaments, chains, jewellery and reproduction statues made at a small scale, and a million other things. These items are not to be exhibited in a personal cabinet of curiosities, but rather used as the *prima materia* for acts of artistic alchemy that metamorphose them into extraordinary three-dimensional, often wall-mounted structures. Kuksi's art is a remarkable synthesis of two artistic practices: the Rococo and the Assemblage. Both notions have their roots in aesthetic rebellion.

The Rococo, the final errant expression of the Baroque period, is highly choreographed and encrusted with asymmetrical ornament. Rococo was primarily a French invention from 1720 to 1730 and is inspired by natural forms. Its name comes from '*rocaille*' – 'rock' and 'broken shell' in English, with acanthus leaves also a significant motif. Rococo forms often exploit the curves found in 'S' or 'C', contorting and embellishing them into extravagant decoration that seems to eschew the simple rules of previous styles and eras. English artist and writer William Hogarth tried to define this new style in relation to these serpentine curves in his *Analysis of Beauty* (1753).[3] He argued that beauty was dependent on six parameters: fitness, variety, regularity, simplicity, intricacy and quality. Intricacy

Kris Kuksi,
Conquest,
2016

The sculptures have the curling organic forms that we are familiar with from the Rococo, but also a steampunk aesthetic not bounded by any time frame – the mechanical mixes with the classical.

is perhaps the most interesting concept particularly when considering Kuksi's work. By 'intricacy', Hogarth means the mental pleasure of the restless eye trying to pursue, comprehend and experience a complete, but complex object – like looking into the leafy canopy of a tree, indefinable yet beautifully infinite. Looking at Kuksi's sculptures has this same effect on the viewer, their intricacy intensely pleasing to the mind and eye.

A Brief History of Subversion
The second rebellious art practice applied by Kuksi – Assemblage – is a 20th-century phenomenon that emerged around 1910 when Pablo Picasso and Georges Braque started to collage found material, for example newspaper text into their Cubist paintings and drawings. Collage was appropriated by the Dadaists as the ideal weapon to aid their subversive commentary on the madness of the First World War and the figures and institutions that facilitated it. In the early 1920s, Dada artist Kurt Schwitters started putting objects together in his *Merz* series. Dadaist collage/assemblage was itself assimilated into the Surrealist movement in the mid-1920s and 1930s.

   Two events and a Surrealist game come to mind, all of which are forerunners that have bearing on Kuksi's compositional lexicon. In 1934, German Surrealist Max Ernst published *Une semaine de bonté*,[4] a collage novel, its raw material salvaged from Victorian pamphlets and catalogues. Through the book's pages we are invited to look into a strange domestic world where normality is turned on its head. Animals and humans are hybridised, objects become animate and their normal scale corrupted and desires unleashed. 'Demure ladies turn into lascivious harlots, or their chastity and virtue is left in tatters as they are subjected to all manner of perils and improprieties. Women are more often victims than vamps, and death, murder, sex and nudity abound. This is a world out of control.'[5]

Kris Kuksi,
Victory of Perseus,
2017

Kuksi's compositional techniques have antecedents that include the subversive Dada collage-makers and the Surrealists. Disparate scales of objects are brought together to create a heady, fecund carnival of elements.

Another Surrealist event that seems relevant to Kuksi's oeuvre occurred at the 'Exposition Internationale du Surréalisme' at the Galerie des Beaux-Arts in Paris in 1938. Visitors to the gallery at 140 rue du Faubourg Saint-Honoré were ushered through a boulevard of 16 female mannequins, each artist responsible for decorating their dummy with Surrealist regalia: chimeric motifs and strange out-of-place objects – spoons, birds, sharks' jaws and a variety of hats – inhabited the bodies, often used to sexualise the mannequins and give them an air of streetwalking provocation.

Finally, Exquisite Corpse is a Surrealist game that named itself; it involves sides of folded paper, on which several people contribute a phrase or drawing, none of the participants having any idea of the nature of the preceding contribution or contributions. The classic example, which gave its name to the game, is the first phrase obtained in this manner: 'THE EXQUISITE CORPSE SHALL DRINK THE YOUNG WINE'. Extraordinary, composite and chimeric bodies/figures could be composed in this way.

British urban-fantasy-fiction writer China Miéville is adept at imagining and describing surreal alternative futures. One of his books in particular chimes with Kuksi's art. *The Last Days of New Paris* (2016)[6] is set in Marseille in 1941 and German-occupied Paris in 1950. In Marseille, a character, on his way to Prague to resurrect a golem as a weapon against the Nazis, befriends a group of Surrealists waiting to go by ship to exile in New York. Somehow, though, a mixture of machinic and occult wizardry and an explosive accident liberate Surrealist art from inertness and make it come alive. Later, wolf tables, furred spoons, a winged monkey with owls' eyes and, of course, exquisite corpses cause deadly havoc and interrupt the battling between the Nazis and the French Resistance, treating each side with similar violence. At the beginning of this article, Miéville describes an exquisite corpse that has been enlivened. It is a direct

> Animals and humans are hybridised, objects become animate and their normal scale corrupted and desires unleashed

description of one collaged by André Breton, Jacqueline Lamba and Yves Tanguy in 1938 that evokes the often-uncanny aspect of Surrealist art.

Prosperity and Fortune
A description of a Kuksi work would be similarly surreal. The central portion of his *Prosperity* (2020), for example, has the backdrop of a bank building. Architecturally, it is of a stripped-back classicism with unfamiliar embellishments such as golden capitals from an unknown style. Behind the building is a festival of 'S' and 'C' curlicues, some reminiscent of octopus tentacles. Atop the bank's roof is a mounted horseman (a St George-type character) about to prod a bull with a lance, one suspects to encourage it and the financial market it represents into ever higher activity and profit. A bear (the symbol of a falling market) approaches the horse and rider unbeknownst from behind. A dollar symbol takes the architectural pride of place on the centre of the bank's portico. A man, maybe about to jump, occupies the corner of an upper cornice. Below, the mythical Fortuna (goddess of fortune and capriciousness) lies dead or unconscious, her black angel wings squashed under her. Her body strewn with coins, she is also used as a landscape for smaller figures acting out other narratives of financial turbulence; greed and poverty, one imagines.

The whole is placed on an exuberant undercroft, its upside-down towers and central truncated obelisk penetrating into an imagined ground. This description does not do the work justice and fails to get over the sheer complexity of Kuksi's sculptures, but it gives a taste of their intensity and semiotic fecundity. Kuksi gives clues to the meaning of the work, but ultimately leaves the viewer space to hypothesise what each piece in the composition might mean and why it is there arranged in juxtaposition with the other pieces.

Kuksi's work can be seen as a series of parallel texts completed and composed

Kris Kuksi, *Prosperity* (detail), 2020

above: The attention to detail in the work is awe-inspiring. It is an art of maximalisation, too complicated for the eye to comprehend in one look; it keeps the eye moving – a very pleasurable activity.

below: (Detail) Much of Kuksi's work includes signs of nascent violence with the frisson of sexuality combined with a blurring of the boundaries between church and state, capital and spectacle.

… protocols of chance, disturbed context, hybrid formal juxtapositions and variable meanings are writ large in Kuksi's sculptures

(Detail) Many of the wall-mounted pieces take place on elaborate, finialled plinths and undercrofts. They take the form of Baroque or Gothic vectors that are composed to provide thrusting, pointing stalactites.

in cahoots with the viewer. His crafting of combination after combination of objects into kaleidoscopic portmanteaux imbued with poiesis, desire and humour provide a cornucopia of new images and concepts. One is reminded, in this respect, of Roger Cardinal's definition of the Surrealist city, and it applies perfectly to Kuksi's oeuvre. Cardinal maintained that the Surrealist city was simultaneously love affair, psychic labyrinth, poetic text, a system of signs, a dream and a palimpsest.[7] These protocols of chance, disturbed context, hybrid formal juxtapositions and variable meanings are writ large in Kuksi's sculptures. Many hands, with many disparate aspirations, make our cities at many scales of operation, perceived and acted within by a myriad of contributors. Some of these actors know about the city's semiotics, but the majority do not. By collecting and repurposing the objects of the everyday, Kuksi allows many unknown designers to influence his work.

Kuksi's sculptures are also time machines, each stretching backwards and forwards, their chronologies simultaneously converging on the retina. The pieces bombard the eye and mind with differing times and differing historical and philosophical understandings of humanity's place in the cosmos.

Kuksi's surreal spatial aspirations, his mnemonic sensibilities and interest in the desiring, fractured Surrealist readings of combinatorial objects give his art great resonance. Such surreal spatial protocols have continued to underpin many of the most interesting architectural and artistic works of the last 100 years and this is still evolving. As curator Ingrid Schaffner has written: 'Working outside Breton's jurisprudence, David Lynch's ant's-eye-view, Angela Carter's violet pornography, Bob Dylan's tombstone blues, ... could also be called Surrealist.'[8] Kuksi fits perfectly into this distinguished cohort of artists. ᴆ

Notes
1. China Miéville, *The Last Days of New Paris*, Picador (London), 2016, p 64.
2. André Breton, 'Manifesto of Surrealism' [1924], in André Breton, *Manifestoes of Surrealism*, trans Richard Seaver and Helen R Lane, University of Michigan (Ann Arbor, MI), 1972, p 14.
3. William Hogarth, *The Analysis of Beauty: written with a view of fixing the fluctuating ideas of taste*, printed by John Reeves (London), 1753.
4. Max Ernst, *Une semaine de bonté*, Editions Jeanne Bucher (Paris), 1934.
5. Neil Spiller, *Architecture and Surrealism: A Blistering Romance*, Thames & Hudson (London), 2016, p 80.
6. Miéville, *op cit*.
7. Roger Cardinal, 'The Soluble City: The Surrealist Perception of Paris', in Dalibor Vesely (ed), ᴆ *Surrealism and Architecture*, 48 (2–3), 1978, pp 143–9.
8. Ingrid Schaffner, *The Return of the Cadavre Exquis*, The Drawing Centre (New York), 1993, p 45.

Text © 2021 John Wiley & Sons Ltd. Images: p 128(t) © Robbie Munn; pp 128(b), 129–33 © Kris Kuksi

CONTRIBUTORS

Phil Ayres is an architect, researcher and educator. He is an associate professor at the Centre for Information Technology and Architecture (CITA) in Copenhagen, which he joined in 2009 after a decade of teaching and research at the Bartlett School of Architecture, University College London (UCL). His research focuses on the design and production of novel bio-hybrid architectural systems that couple technical and living complexes, together with the development of complementary design frameworks. He has pursued this research in the context of the EU-funded projects Flora Robotica and Fungal Architectures. He is also the editor of the title *Persistent Modelling: Extending the Role of Architectural Representation* (Routledge, 2012).

Kathy Battista is an art historian and a curator of exhibitions in museums, galleries and non-profits. Her research is primarily focused on cross-generational feminist art, in particular performance and body-oriented practice. Other research topics include the intersection of gender, art and fashion, and the art market. She has authored numerous books including *New York New Wave: The Legacy of Feminist Art in Emerging Practice* (IB Tauris, 2015) and *Renegotiating the Body: Feminist Art in 1970s London* (IB Tauris, 2012). She is also co-editor (with Bryan Faller) of *Creative Legacies: Critical Issues in Artist Estates* (Lund Humphries, 2020), a book on artists' estates and foundations.

Thea Brejzek is Professor of Spatial Theory at the University of Technology Sydney. She publishes and lectures widely on the history and theory of scenography and performative environments with a particular interest in transdisciplinary practices, and in 2011 was the Founding Curator for Theory at the Prague Quadrennial for Performance Design and Space. She is a member of the scientific advisory board of the Bauhaus Dessau and Associate Editor of the Routledge journal *Theatre and Performance Design*. Recent publications include *The Model as Performance: Staging Space in Theatre and Architecture* (Bloomsbury, 2018), co-authored with Lawrence Wallen.

Pascal Bronner graduated from the Bartlett School of Architecture, UCL in 2009, and was awarded the RIBA Bronze Medal Commendation, the Fitzroy Drawing Prize, the Serjeant Award for Excellence in Drawing and the Sir Bannister Fletcher Medal. He currently runs both a Master's and BSc unit, alongside Thomas Hillier, at the University of Greenwich in London and at the Bartlett. His work has been exhibited around the world, including at the Dallas Center for Architecture, the AIA Headquarters in Washington DC, and at the Royal Academy of Arts Summer Exhibition in London. In 2010 he was invited to exhibit his project *New Malacovia* as part of 'London 8', an exhibition at the Southern California Institute of Architecture (SCI-Arc) in Los Angeles curated by Sir Peter Cook. In 2012 he co-founded FleaFollyArchitects with Hillier.

Mark Cousins was the Head of History and Theory Studies at the Architectural Association (AA) in London. Educated at the University of Oxford and the Warburg Institute, he taught across the undergraduate, postgraduate and PhD programmes at the AA. His Friday-evening public lectures, which ran for over 30 years, were widely influential. Zaha Hadid sought Cousins's advice in planning the MAXXI – National Museum of 21st-Century Art in Rome. Cousins published numerous essays and one book, *Michel Foucault* (Palgrave Macmillan, 1984) with Athar Hussein. He was a founding member of the London Consortium, and a Visiting Professor at Columbia University in New York and Southeast University in Nanjing, China.

James A Craig is Lecturer in Architecture at Newcastle University. Prior to this he worked as a visiting lecturer at the University of Greenwich and at University College Cork, and as a Studio Master at the AA. He holds an AHRC Northern Bridge-funded scholarship (awarded 2019) for his ongoing creative practice doctoral studies that follow the title: 'The Autobiographical Hinge: Revealing the Intermediate Area of Experience in Architectural Representation'. At the centre of his research is a query over the capacity for representational objects to expose the intersubjective relationships between architects and their tools.

Kate Davies is an artist and architect. Her work explores the complexities of contemporary landscape, from landscapes of extraction, manufacturing and logistics to remote territories, wilderness and ancestral homelands. She is particularly interested in the places where conflicting value systems coexist. She is co-founder of the Unknown Fields travelling design studio, and art and architecture collective Liquid Factory. Her work has been exhibited internationally, showcased by mainstream media, and is held in the permanent collections of major museums. She lectures and teaches design studios in architecture, new media and landscape at the AA and the Bartlett School of Architecture

Ryan Dillon is the Head of Communications at the AA, has been Unit Master of Intermediate 5 at the school since 2013, and currently teaches in the History and Theory and Design Research Laboratory (DRL) programmes. He previously worked at Moshe Safdie Architects on projects such as the Khalsa Heritage Complex and the Peabody Essex Museum. He studied architecture at Syracuse University in New York, and holds an MA(Dist) from the AA's Histories and Theories programme. His research explores the Greenwich Meridian's role in establishing time as a constraint within everyday social routines, the built environment and capitalist societies.

Thomas Hillier graduated from the Bartlett School of Architecture in 2008, and was awarded the Dean's List for Excellence in Design. He was nominated as the Bartlett's best graduating student for *Building Design* magazine's 'Best of 2008' awards. He currently runs both a Master's and BSc unit, alongside Pascal Bronner, at the University of Greenwich and the Bartlett. His work has been published extensively in books and journals such as *Blueprint*, *Icon* and *Domus*. He has exhibited around the world, notably in 2012 with his first solo exhibition at the RAW Gallery of Architecture in Winnipeg, Canada. In 2010 his project The Migration of Mel & Judith was exhibited at the Royal Academy of Arts Summer Exhibition in London.

Christian Hubert is an architectural designer, critic and translator based in New York. He holds a Bachelor of Arts from Columbia University, and a MArch degree from Harvard University. During the 1980s and 1990s he was affiliated with the Institute for Architecture and Urban Studies (IAUS) in New York, where he was elected to the Fellowship. He has designed residences for artists, collectors and scholars, as well as galleries and museum installations, including 'The American Century' (1999–2000) exhibition at the Whitney Museum of American Art in New York. His published essays include 'The Ruins of Representation', 'Playtime', and a hypertext dictionary of critical terms, posted online at: www.christianhubert.com/writing.

Matt Ozga-Lawn is Lecturer in Architecture at Newcastle University. Along with James A Craig, he runs the experimental design platform Stasus, which was published in *Pamphlet Architecture 32: Resilience* (Princeton Architectural Press, 2012). He teaches in the architecture department at both undergraduate and final year (MArch stage 6) level. He coordinates the innovative module Theory into Practice, which focuses on representational techniques and the design process.

Chad Randl is the Art DeMuro Assistant Professor in the Historic Preservation programme at the University of Oregon in Eugene. He is the author of *A-Frame* (2004) and *Revolving Architecture: A History of Buildings that Rotate, Swivel and Pivot* (2008), both published by Princeton Architectural Press. His research explores cultures of building design and inhabitation focusing on change, recreation and popular taste. His work has appeared in *Buildings & Landscapes*, *The Senses and Society*, the *Journal of Architecture* and the edited volume *Archi.Pop*. He is currently writing a history of home improvement in the US.

Neil Spiller is Editor of ⌀, and was previously Hawksmoor Chair of Architecture and Landscape and Deputy Pro Vice Chancellor at the University of Greenwich. Prior to this he was Vice Dean at the Bartlett School of Architecture. He has made an international reputation as an architect, designer, artist, teacher, writer and polemicist. He is the founding director of the Advanced Virtual and Technological Architecture Research (AVATAR) group, which continues to push the boundaries of architectural design and discourse in the face of the impact of 21st-century technologies. Its current preoccupations include augmented and mixed realities and other metamorphic technologies.

Theodore Spyropoulos is an artist, architect and educator. He is the Director of the AA Design Research Lab (AADRL) and resident artist at Somerset House in London. He has chaired the AA Graduate School, been Professor of Architecture at the Staedelschule, Frankfurt, and a visiting research fellow at the Massachusetts Institute of Technology (MIT) Center for Advanced Visual Studies. He directs the experimental art, architecture and design practice Minimaforms. His work has been acquired by international collections including the FRAC Centre, Orléans, France. He has exhibited at MoMA, the Onassis Cultural Centre in Athens, Barbican Centre and Institute of Contemporary Arts (ICA) in London, and Detroit Institute of Arts.

Lawrence Wallen is an architect, visual artist and professor at the School of Architecture, University of Technology Sydney. His practice operates across a range of modes employing scenographic strategies and performative readings resulting in an extensive collection of books, drawings, performances and architectures. He is co-author (with Thea Brejzek) of *The Model as Performance* (Bloomsbury, 2018), and author of the essay 'On the Reconstruction of Landscape', published in the journal *Leonardo* in February 2020. Recent exhibitions in Vilnius, France, Cyprus, Sydney, Cairo and Italy complemented by residencies and fellowships have operated in dialogue with and underpinned his writing.

Mark JP Wolf is a Full Professor in the Communication Department at Concordia University Wisconsin in Mequon. His 23 books include *Building Imaginary Worlds: The Theory and History of Subcreation* (2012), *Revisiting Imaginary Worlds: A Subcreation Studies Anthology* (2017), *The Routledge Companion to Imaginary Worlds* (2017), *World-Builders on World-Building: An Exploration of Subcreation* (2020) and *Exploring Imaginary Worlds: Essays on Media, Structure, and Subcreation* (2021), all published by Routledge.

What is *Architectural Design*?

Founded in 1930, *Architectural Design* (△) is an influential and prestigious publication. It combines the currency and topicality of a newsstand journal with the rigour and production qualities of a book. With an almost unrivalled reputation worldwide, it is consistently at the forefront of cultural thought and design.

Issues of △ are edited either by the journal Editor, Neil Spiller, or by an invited Guest-Editor. Renowned for being at the leading edge of design and new technologies, △ also covers themes as diverse as architectural history, the environment, interior design, landscape architecture and urban design.

Provocative and pioneering, △ inspires theoretical, creative and technological advances. It questions the outcome of technical innovations as well as the far-reaching social, cultural and environmental challenges that present themselves today.

For further information on △, subscriptions and purchasing single issues see:

https://onlinelibrary.wiley.com/journal/15542769

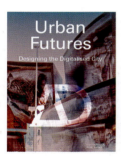
Volume 90 No 3
ISBN 978 1119 617563

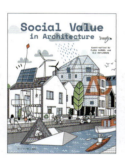
Volume 90 No 4
ISBN 978 1119 576440

Volume 90 No 5
ISBN 978 1119 651581

Volume 90 No 6
ISBN 978 1119 685271

Volume 91 No 1
ISBN 978 1119 717669

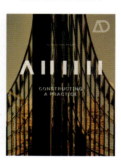
Volume 91 No 2
ISBN 978 1119 717485

Individual backlist issues of △ are available as books for purchase starting at £29.99 / US$45.00

www.wiley.com

How to Subscribe
With 6 issues a year, you can subscribe to △ (either print, online or through the △ App for iPad)

Institutional subscription
£346 / $646
print or online

Institutional subscription
£433 / $808
combined print and online

Personal-rate subscription
£146 / $229
print and iPad access

Student-rate subscription
£93 / $147
print only

△ App for iPad
6-issue subscription:
£44.99 / US$64.99
Individual issue:
£9.99 / US$13.99

To subscribe to print or online
E: cs-journals@wiley.com

Americas
E: cs-journals@wiley.com
T: +1 877 762 2974

Europe, Middle East and Africa
E: cs-journals@wiley.com
T: +44 (0) 1865 778315

Asia Pacific
E: cs-journals@wiley.com
T: +65 6511 8000

Japan (for Japanese-speaking support)
E: cs-japan@wiley.com
T: +65 6511 8010

Visit our Online Customer Help
available in 7 languages at www.wileycustomerhelp.com/ask